I0796818

Praise for *The Self-Fed Farm and Garden*

"Over the years, Eliot Coleman has been a mentor and guiding light in my own journey as a market gardener. In *The Self-Fed Farm and Garden*, he goes even deeper—offering a radical yet timeless understanding of what organic farming is truly about. A true contrarian in the best sense, only Eliot could bring such clarity and wisdom to the conversation. He reminds us that the farm is not just a place of production, but a space of ecological harmony and farmer autonomy. This book is the culmination of a lifetime of insight and a must-read for anyone who believes that food can be a force for change. We've all been waiting for a book like this—a new classic for the ages."

—Jean-Martin Fortier, author of *The Market Gardener*; founder, the Market Gardener Institute

"In his unparalleled professor-meets-practice persona, Eliot Coleman brings 'I can do this' inspiration to farmers and gardeners yearning to break free from off-farm fertility dependency. Few ideas could be more needed and prescient than soil fertility self-reliance. *The Self-Fed Farm and Garden* explains how, and in so doing, this little book spells economic and process freedom."

—Joel Salatin, farmer; author; food freedom advocate

"Generations of growers continue to be inspired by Eliot Coleman's organic vision, practical methods, and articulate instruction. *The Self-Fed Farm and Garden* challenges its readers to reconsider the restorative power of plants, healthy soil, and the integrity of organic farming in the modern world."

—Jack Algiere, Stone Barns Center

"What happens when you combine one of the most talented organic growers alive with a scholar who writes beautifully? You get Eliot Coleman's new classic, *The Self-Fed Farm and Garden*. I have been a student of Eliot's for many years, and he continues to challenge my thinking and change my life. This book steps back from the frantic energy of the marketplace to take a second look at what we mean when we say 'organic,' why we need it, and how to do it right. Eliot helped me to learn all this in the first place.

His remembrance of things past and hope for things yet to come is needed. May many read this book and *act* on it. It will change you and the world. This is the book we have needed."

—Dave Chapman, founder and codirector, Real Organic Project

"Leave it to America's foremost organic farmer, agricultural historian, and talented teacher to map out the final frontier for organic farmers and gardeners. In this book about Big Biomass, Eliot Coleman reveals his practical system of fertility independence. He explains that we all can grow our own crop fertility amid the process of rotating our fields and gardens. The distillation is to yearly dedicate a third of your ground for a planting of lush, soil-nourishing crops of winter rye and hairy vetch. In this era of ubiquitous pollution, Eliot's *Self-Fed* solution is safe, sound, and timely."

—Jim Gerritsen, founder and farmer, Wood Prairie Family Farm

"Eliot Coleman's *The Self-Fed Farm and Garden* is a treasury of his stored experience, gleaned over a lifetime of curious experimentation, reflection, and practice dropping seed and gleaning his harvest. For Eliot, self-reliance and discovering the beauty of organic practice are principles that frame his experience. The book is a journal of his lessons learned: timeless truths bound by the elegance of what nature reveals to the observant eye. His chronicles can be appreciated by all who choose to produce food at any scale. The book is also a treat to read, allowing us insight into a lifetime of learning by a curious and principled farmer. It ensures that future generations of new farmers, and old farmers like myself, will utilize his observations to make their farms more elegant and wildly more productive—places that reflect their own principled assembly of the pieces that fit their place, climate, community, and mind."

—Paul Muller, cofounder and farmer, Full Belly Farm

"Eliot Coleman's ideal, *the self-fed farm*, is regenerative, resilient, and resourceful. It relies on the regenerative capacity of diverse living ecosystems to make the best use of Nature's bountiful resources. It is as close to *sustainable farming* as it's possible to get in today's unsustainable society."

—John Ikerd, professor emeritus of agricultural economics, University of Missouri-Columbia

"Eliot Coleman evokes a self-sustaining concept of biological completeness to elegantly explain how to create a self-fed farm by cycling and recycling on-farm fertility, addressing soil, plant, and farm needs while providing nutritious food."

—KRIS NICHOLS, lead soil scientist, Food Water Wellness Foundation; former chief scientist, Rodale Institute

Praise for Eliot Coleman

"The incomparable Eliot Coleman . . . serious, meticulous, inspiring."

—*THE NEW YORK TIMES*

"One of America's most innovative farmers."

—MICHAEL POLLAN

"I have been a devotee of Eliot's for years, fully agreeing with his methods for growing—in winter, spring, summer, and fall—tasty, nutritious produce with a minimum consumption of fossil fuels.

"I know of no other person . . . who can produce better results on the land with an economy of effort and means than Eliot. He has transformed gardening from a task to a craft, and finally to what Stewart Brand would call 'local science.'"

—PAUL HAWKEN, author of *Blessed Unrest*; editor of *Drawdown*

"Coleman tracks his own constant search for perfection, a quality that has led more than one young farmer to exclaim, 'I'd follow him anywhere.'"

—JOAN DYE GUSSOW, author of *This Organic Life*

"Coleman speaks as if to a fellow home or market gardener, sharing what works for him and discussing what he knows and what he doesn't know."

—*THE SAN FRANCISCO CHRONICLE*

Also by Eliot Coleman

The New Organic Grower:
A Master's Manual of Tools and Techniques
for the Home and Market Gardener

Four-Season Harvest:
Organic Vegetables from
Your Home Garden All Year Long

The Winter Harvest Handbook:
Year-Round Vegetable Production Using
Deep-Organic Techniques and
Unheated Greenhouses

The SELF-FED FARM *and* GARDEN

A Return to the Roots of the Organic Method

ELIOT COLEMAN

Introduction by DAN BARBER

Chelsea Green Publishing
White River Junction, Vermont
London, UK

First published in 2025 by Chelsea Green Publishing | PO Box 4529 | White River Junction, VT 05001 | West Wing, Somerset House, Strand | London, WC2R 1LA, UK | www.chelseagreen.com
A Division of Rizzoli International Publications, Inc. | 49 West 27th Street | New York, NY 10001 | www.rizzoliusa.com

Publisher: Charles Miers
Deputy Publisher: Matthew Derr
Project Manager: Natalie Wallace
Developmental Editor: Fern Marshall Bradley
Associate Editor: Amalia Herren-Lage
Copy Editor: Diane Durrett
Proofreader: Angela Boyle
Indexer: Linda Hallinger
Designer: Melissa Jacobson

ISBN 978-1-64502-306-7 (hardcover) | ISBN 978-1-64502-307-4 (ebook) | ISBN 978-1-64502-308-1 (audiobook)
Library of Congress Control Number: 2025028322 (print) | 2025028323 (ebook)

Our Commitment to Green Publishing
Chelsea Green sees publishing as a tool for cultural change and ecological stewardship. We strive to align our book manufacturing practices with our editorial mission and to reduce the impact of our business enterprise in the environment. We print our books using vegetable-based inks whenever possible. This book may cost slightly more because it was printed on paper supplied by Versa from well-managed, FSC®-certified forests and other controlled sources.

Authorized EU representative for product safety and compliance
Mondadori Libri S.p.A. | www.mondadori.it
via Gian Battista Vico 42 | Milan, Italy 20123

Printed in the United States of America.
10 9 8 7 6 5 4 3 2 1 25 26 27 28 29

To the many friends who
read this manuscript in progress
and sent me useful feedback,
with my heartfelt thanks.

Every well-managed soil should grow richer rather than poorer; and, speaking broadly, the farm should have within itself the power of perpetuating itself.

LIBERTY HYDE BAILEY,
The State and the Farmer (1908)

The soil in which plants grow must not be thought of as a layer of dead matter in which plant roots are fastened . . . but rather as a very busy biological laboratory.

ADRIAN PIETERS,
Green Manuring (1927)

It should not be necessary to bring anything on to the farm from outside in the nature of manure or foodstuffs. That is the essential difference between a farm and a factory.

HUGH MAUDE,
The Farm (1943)

Vegetation adds to soil fertility, and there is no better way to husband a piece of land so well as to let something grow on it and return its residue to the soil.

PETER FARB,
Living Earth (1959)

CONTENTS

PREFACE

Back to the Roots

When the organic growing movement began many years ago, the pioneering farmers created and maintained their fertile soils by partnering with the earth's resources. They sowed green manures and cover crops and shallowly tilled them into the soil, they grew legumes to add nitrogen, they devised effective crop rotation systems for control of plant diseases and weeds, they spread rock powders as natural mineral sources, they made soil-nourishing compost from locally available, pure organic wastes, they used manure from their own livestock, and so forth. They created an exceptionally clean, safe, self-contained production system. Nowadays it is a much different world. Those traditional, farmer-managed, soil-improving *processes* have been replaced by *products*—purchased products from off the farm.

My greatest concern is about the questionable quality of the compost products available for purchase today. Compost has always had a reputation for goodness among organic growers. But what about the unavoidable residual environmental contamination of the ingredients from which the numerous municipal-waste and confined-livestock composts marketed today are made—contaminants such as pesticides, herbicides, antibiotics, veterinary drugs, hormones, heavy metals, and the like? Purchased inputs of industrially suspect manure from confined animal feeding operations (CAFOs) or purchased compost made from the municipal waste stream have become just as much of a crutch for today's organic vegetable growers as chemical fertilizers are crutches for chemical farms. Unfortunately, those purchased organic-in-name-only inputs no longer meet the traditional pure standards for inputs always associated with organic farming. The manufacturers of commercial composts are paid to dispose of waste products. Most of those composts, which are enthusiastically used by many organic vegetable growers, are not the high-quality compost celebrated by Sir Albert Howard, the renowned nineteenth century botanist and sustainable agriculture pioneer. The outward appearance of the finished compost doesn't give any clues as to what was disposed of to make it.

One of the principal motivations for Four Season Farm's preference for organic soil care has been to provide our customers with food free of the chemical pollution and the industrial toxins that pervade our world. Because the only sure way to have both a clean as well as a fertile soil is to grow our own soil organic matter right on the farm, that is what we do. The aim of our "self-fed farm" project is to consciously separate our farm's practices from the industrially influenced, inadequate shortcuts that so many organic farmers now rely on. The aim of this book is to detail the simple and productive methods we use to achieve our goal of producing clean and nourishing food in perpetuity.

Preface

FOUR SEASON FARM
REAL FARMING · REAL FOOD
FARMSTAND OPEN
~ SATURDAYS 8 - 12 ~
REAL
ORGANIC
PROJECT

A statement titled "Organic Farming as Practiced by Four Season Farm" is proudly posted at our farmstand. It reads as follows:

The popular press defines organic farming by its rejection of chemicals.

A better portrayal defines organic farming by its embrace of the soil's biological systems.

1. Classical organic farming is based upon the creation and maintenance of a biologically active fertile soil.
2. Organic farming succeeds because of the benefits derived from that soil. Organic farmers focus on correcting the cause of pest problems (weak plants) by strengthening the plant through optimum soil conditions, rather than merely treating the symptom (pest damage) by trying to kill the pests that are attracted to weak plants. The production of pest-free plants and livestock with active immune systems is a direct outcome of organic soil care that, as scientific studies have consistently shown, induces pest and disease resistance.
3. Research into the marvelously complex soil microbiome (and the associated human microbiome) is revealing the vital microbiological truths that link both soil health and human health. That research underscores the intuitive brilliance of the founding organic farmers.
4. As a bonus, fertile soil produces food of the highest nutritional quality. That was one of the foremost initial aims of the soil care techniques that became organic farming.

5. Long-term soil productivity does not require frequent input of purchased fertilizer inputs from off the farm. It can be created and maintained by the use of farm-made compost, crop rotations, green manures, cover crops, nitrogen-fixing legumes, grazing livestock, shallow cultivation, nutrient-dense powdered rock, enhanced biodiversity, and other time-honored practices that nurture the boundless energy and logic of the earth. Organic farming is a circle of endless renewal and it will succeed wherever there is soil.
6. The inclusion of deep-rooting forbs in rotationally grazed grass/legume pastures within the rotation helps to maintain fertility and make available the almost inexhaustible mineral supply from the lower levels of the soil. Grazing livestock benefit the soil; diverse pasture benefits the livestock.
7. Most significant of all, since soil fertility on the classical organic farm is not powered by purchased inputs from outside the farm but, rather, by easily understood and universally applicable soil management practices conducted within the farm, this food production system is accessible at low cost by farmers everywhere and can thus nourish the planet with exceptional food in perpetuity. As a further benefit, the self-fed organic farm avoids the accidental introduction of industrial contaminants from outside the farm which is a perpetual threat to the purity and integrity of input-dependent systems.

INTRODUCTION

I first learned about plant health from Eliot Coleman when I was a college senior, deep in the throes of final exams—which is to say, I was procrastinating. Instead of studying, I wandered the library's stacks, flipped through periodicals, made trips to the vending machine, and browsed the latest arrivals laid out on the wooden table by the entrance. That's when I first spotted *The New Organic Grower*. Its author, Eliot Coleman, looked out from the cover with the kind of blue-eyed conviction that suggested he knew something the rest of us didn't.

He did. Eliot's idea—that healthy plants grown in healthy soil are naturally resistant to pests—arrived at the perfect moment for a college senior with vague anarchist sympathies (really, just a liberal arts major) hungry for ideas that were both radical and elegantly simple.

Feed the soil, and the tiny creatures living in it, with care and attention, and pests will almost always be incapable of inflicting damage.

I didn't yet realize that Eliot's dirt-first dogma would become the foundation for my own cooking. Chemical farming, I decided, was less a necessity than an elaborate misunderstanding. "Truth to power," would have been something I muttered to myself, delighted by a worldview that turned conventional wisdom on its head.

Eliot, for his part, is fond of saying that the microbes in the soil should be fed first—let them have first pick at the buffet, and they'll take care of the rest. It's an ecological "if you build it, they will come"—except that if you feed the soil, the good microbes and nutrients will stay and, eventually, the pests will leave.

My radicalism tempered with age, but a decade after that first encounter with Eliot's work, I had the good sense to reach out to him for help in designing the farm at Stone Barns Center for Food & Agriculture, which would become the home of my restaurant, Blue Hill. Eliot's task was to identify the best land for growing strong, healthy vegetables without chemical intervention.

I've told the story of his first visit to the site more times than I can count, because that crisp November afternoon in 2002 marked yet another turning point in my life as a chef. As the light faded on that cold, slow-burning day in the time of year best known for its impression of perpetual dusk, Eliot walked ahead of me, surveying the fields. He'd initially settled on a flat, healthy stretch of land, but then we came to a long, upward-sloping field beside the largest of the stone barns, once the heart of the property's dairy operation. Eliot paused, scanning the six-acre expanse. "The cows probably pastured here," he mused.

Then, in a flash, he dropped his bag and, in a move more reminiscent of a border collie than a consultant, started running—darting across the old pasture, weaving between rocks and thistles, head swiveling to track the setting sun. He passed what would later become rows of tomatoes, cucumbers, fava beans, and parsnips, finally reaching the highest point, where he raised his finger to the sky. Then he was off again, making his way to the northeast corner,

Introduction

stopping at last, hands on hips, to study the land. Even in his early sixties (and still today), there was something wild and hopeful in his movements—curious, observant, energized by a deep, intuitive connection to nature, and filled with the hopeful energy of someone who finds joy in the possibility that the best answer is not the obvious one. I watched in awe.

"Hey, freaking cool," he grinned as he returned, his blond hair gleaming in the fading light, eyes wide and sparkling. "This is the field. Forget the other one. I'm getting hungry just thinking about what you'll grow here." I asked if he'd changed his mind based on the angle of the sun. "The sun? Oh, hell, no. I was just admiring it. No, I wanted to be sure this had been pasture for the cows."

Eliot explained that fields closest to the barns were usually the most heavily grazed—and thus the most richly fertilized by manure. In this case, he was right; we later confirmed that this pasture had once been grazed by the Rockefeller family's dairy herd.

He scooped up a fistful of soil and turned his hand so I could see: "Black enough for you?" he laughed, marveling at its richness. In the dying light, I stared at his handful of earth, realizing that the true recipe for great cooking isn't a recipe—it's an idea.

The idea is to honor the astonishing network of microscopic life in the soil—the full periodic table of elements, offered up if you let it.

Until then, I'd held on to two simple misconceptions: that chemical farming kills soil by poisoning it (which it can), and that eating those chemicals is unappetizing and harmful (which it probably is). But both miss the chef's larger point: chemical farming starves the soil's complex community, depriving it of anything good to eat—and, in doing so, robs soil of its natural resilience and food of its flavor.

"Why limit the hand that feeds you?" Eliot asked, reading my mind. "Thinking we can substitute a few soluble elements for a whole living system is like believing an intravenous needle could deliver a delicious meal."

What a pleasure, then, to introduce this new book and find Eliot's ideas as vital and radical as ever, now distilled into a practical guide for anyone who wants to feed the farm that feeds them.

In this sense, what you're holding is a pantry cookbook—and I mean that as the greatest compliment. Eliot's blueprint for nourishing the soil with what's already on hand is as accessible for every home gardener and farmer as it is essential.

That Eliot Coleman has continued, decade after decade, to deepen his thinking, remaining somehow both radical and accessible, is a testament to his curiosity, his generosity, and his refusal to accept the world as it is handed to him. May these pages do for you what Eliot once did for me: prompt you to look down, dig in, and realize that the revolution is still possible, and that, because of it, our food will taste good.

DAN BARBER, Blue Hill at Stone Barns
July 2025

CHAPTER ONE

Celebrating Life

A fertile soil is filled with life. A single handful of fertile soil is said to contain considerably more living creatures than the human population of planet Earth. Traditional organic farming is based upon supporting the life in that fertile soil by adding organic matter to nourish all those living soil creatures, whose activities—most importantly the decomposition of that organic matter—result in the creation of plant food. In this book I suggest that commercial organic vegetable growers can enjoy exceptionally fertile growing conditions, a guaranteed clean (unpolluted) soil, and far lower expenses by growing that indispensable organic matter themselves—by which I mean extensive use of green manures and cover crops as the "inputs" for soil fertility—rather than relying upon purchased, commercial "organic" soil-fertility stimulants from outside their farm. At Four Season Farm we exclusively incorporate a wide range of green manures and cover crops, which include grasses, legumes, and edible forbs, to maintain our farm's soil fertility, year-round. As I have always understood it, the original goal of organic farming was *not* to directly apply purchased fertilizers as plant food. In contrast, such fertilizers were the *sole* focus of chemical agriculture because it had no other option. The original organic goal was to create and maintain a biologically active fertile soil filled with homegrown organic matter and a vigorous population of soil microorganisms, microarthropods, and earthworms able to recycle that organic matter into nutrients for another generation of plants. The ideal of dependable clean food production based upon maximum mobilization of the earth's inherent natural soil fertility potential was the principal aim of the pioneers who developed the earliest organic farms.

Two Principles to Define Organic

Over the years I have collected a considerable library of books written by those early organic adherents who expressed their ideas about the concepts that became *organic farming*. Many times, I have wandered through those tomes comparing one to the other because I am endlessly curious about the deeper thinking of all those who pioneered this approach. In the most basic terms, how did they define *organic* and how did they see organic operating? Is there a clear set of traditional principles whose key positive attributes can be expressed in simple language?

Thanks to those wanderings, I have determined two principles that I think embrace and define an organic farm. The first of these is the clear acknowledgement that organic growing is a biologically based, not a chemically based, agriculture. The natural growth of plants is governed by the activities of those soil-dwelling microorganisms, microarthropods, and earthworms previously mentioned. Those creatures—the key to soil fertility on an organic farm—are alive. One of the earliest books to appreciate that fact, a book I consider to be one of the oldest on the organic bookshelf, is entitled *Essays on Rural Hygiene* and was first published in 1893. Its author, Dr. George Vivian Poore,

Eliot consulting a book in his library.

was not a farmer. He was an English medical doctor and, in addition, a professional sanitarian, which means that he had an early understanding of the importance of soil life for recycling organic wastes in the soil. Dr. Poore's book contains a chapter entitled "The Living Earth" in which he states:

> *We have arrived of late years at a certain knowledge of the fact that the mould which forms the upper stratum of the ground on which we live is teeming with*

> *life and this fact is one of prime importance . . . the black vegetable mould which lies upon the surface of the earth is largely composed of organic matter, which is not to be wondered at, seeing that every organized thing, whether animal or vegetable, which inhabits this globe, falls, when dead, upon the earth and becomes incorporated with it. . . . Enormous numbers of bacteria have always been found in the soil by the most various observers. . . . These micro-organisms of the soil are very active in producing changes in organic matter added to the soil. . . . The oxidation and nitrification of organic matter in the soil is a biological question, pure and simple. It is an effect produced by the* living earth*. . . . There can be no better illustration of the true economy of nature than the action of the microbes in the soil on the conversion of organic matter into soluble salts and gases which serve as food for plants. . . . To keep the soil healthy, to keep up its appetite for dirt [organic matter] and its power of digestion, the only thing necessary is tillage.*[1]

The second principle that I believe defines an organic farm is the concept of the farm as a living organism. This principle found expression some thirty years after Poore's book, when the Austrian philosopher Rudolf Steiner presented his famed agriculture course in June 1924. That course created the foundation for biodynamic farming. Even though Steiner's message was mostly directed toward his fellow anthroposophists, it appears that some seeds found fertile soil among the early organic farmers as well. Both groups at the time were independently looking for whole and complete agricultural philosophies. The living organism concept was eloquently reemphasized fifteen years after Steiner's course by Lord Northbourne (Walter Ernest Christopher James) in his seminal work *Look to the Land* (1940):

> *The higher degree of biological self-sufficiency achieved by a farm, the more alive, the more vigorous, and the more creative it will be. Real liveliness comes from within, not from without; it is the sign of that internal self-sufficiency which is vitality. . . . The soil and the microorganisms in it together with the plants growing on it form an organic whole. . . . The farm itself must have a biological completeness, it must be a living entity. . . . A self-contained organic farm is no mere theoretical dream.*[2]

Those two principles—that organic growing is biologically based and that a farm is a living organism—along with all the biological processes connecting them have influenced my understanding of true organic growing and created the background for the ideas expressed in this book.

My Introduction to Farming

When I first began growing organic vegetables, almost 60 years ago, I was living in a rural New Hampshire village where I had easy access to piles of rotted horse and chicken manure from generous neighbors. In addition, there were nearby sources of finely ground stone-dust wastes from the granite polishing industry, which were valuable for enhancing long-term soil fertility. I made use of all of them, and right from the start my crops were bounteous.

However, I was fully aware that my instant success at growing vegetables organically was not a universal solution for everyone. My beautiful crops were made possible by my easy access to those organic inputs for soil fertility. I was bothered by the fact that I had not been involved in creating those inputs and, therefore, my efforts were not viable, vis-à-vis what I have come to consider the key factor in dependable long-term food production—the ability to create and maintain a highly productive fertile soil from a farm's own resources.

Yes, I had certainly proven that the basic concepts of organic growing were sound and worked incredibly well, even for a rank beginner. But some logical and yet perplexing questions were always in the back of my mind.

- Would I be able to continue producing such exceptional vegetables if my situation changed and I no longer had access to those free manure inputs?
- Would I need to keep livestock?
- Would growing soil-improving green manure crops or making lots of compost from organic wastes on the farm provide adequate soil fertility to replace what I was presently receiving from my neighbors' generosity?
- For how long would the rock powders I added to the soil successfully supplement elements that my soil did not initially contain in quantity?
- For how long would those rock powders release the necessary minerals to replenish what my crops were taking from the soil?
- Could a food producer ever create and maintain a fertile soil simply from internal systems?

- Are nature's resources so extensive that an independent, self-contained farm might be configured to produce food in perpetuity despite selling produce off the farm?

Over many years I have tried to find positive answers to those questions because I have a deep interest in the long-term survival of successful small-scale organic food production.

Four Season Farm's land was cleared from coniferous spruce and fir forest, meaning it had less organic material buildup and was quite sandy. Over time and through addition of lots of compost and green manures, it has become the rich, dark soil seen here.

I made the decision to become a commercial organic market gardener after that first summer's gardening experience. I had been initially inspired to grow food after reading Scott and Helen Nearing's classic book, *Living the Good Life* (1954). The idea of running a successful small organic farm that produced exceptionally nutritious vegetables to feed my neighbors was very appealing to me. In the local library there were many back issues of Rodale's *Organic Gardening* magazine, and I spent the winter reading them. Because magazines are supported far more by advertisements than by subscriptions, the issues were filled with exhortations to buy products—organic fertilizers, organic pesticides, and so forth—that I would supposedly need if I wanted to succeed in organic vegetable growing.

I was eager to acquire more detailed instruction on organic farming beyond what those magazines offered, so I began reading all the books I could find on the subject. The first titles had been published in Europe. I was surprised to find some that dated as far back as the 1890s. They were the foundational works of the nascent organic farming movement. As I read, I soon came to realize that there was an important difference between the organic *farming* that interested me and the organic *gardening* I had read about in the magazines.

Thanks to all those ads in such magazines during the back-to-the-land revival in the 1960s and '70s, much organic gardening today is principally focused on using all those purchased organic fertilizers—bags of bone meal, dried cow manure, soybean meal, blood meal, greensand, and worm castings, not to mention gallons of liquid fish emulsion—to quickly create a temporary soil fertility; fertility, however, that appeared to me to be unnatural and impermanent because it was not connected to the past and present history of that soil. In contrast, the classic farming books

A Concise Glossary

These traditional soil fertility management processes are the foundation of organic farming.

Incorporating compost: Adding partially decomposed plant and animal wastes to the soil for their fertilizing value.

Rotating crops: Changing the choice of crop grown on a given field or garden bed each year to help manage weeds and prevent plant disease.

Using green manures: Sowing a stand of deeply rooted plants specifically to incorporate into the soil to enhance soil fertility.

Sowing cover crops: Sowing crops especially to scavenge soluble plant nutrients from the soil and protect the soil from erosion and temperature extremes over the winter.

Integrating livestock: Collecting manure from the farm's livestock for use as a valuable soil fertility improver; grazing livestock on land planted to pasture crops as part of the overall soil management system.

Growing legumes: Choosing to include crops from the legume family of plants that has the ability to extract nitrogen from the air and add it to the soil.

Replenishing soil organic matter: Adding plant and animal residues that improve soil structure and provide food for soil microorganisms.

Shallow incorporation: Adding organic matter to the soil by mixing it into the top 4 inches (10 cm), which provides the most effective result.

by European authors, many of which soon became part of my personal library, made it quite clear that organic farming was different. To achieve the living, biologically active fertile soil that distinguishes an organic farm from a chemical farm, farmers did not need to base their soil fertility on purchased product inputs. The organic farming books explained that farmers themselves could and should create the foundation of soil fertility on their farms and in their soils by employing a range of universally acknowledged, natural soil fertility–enhancement processes that emphasize the importance of soil organic matter. Those processes include making farm-scale compost, devising effective crop rotations, growing deep-rooting green manures, sowing cover crops for soil protection, integrating livestock where appropriate, growing legumes for soil nitrogen, replenishing soil organic matter, shallowly incorporating organic residues, and so forth. They are what I have come to think of as soil fertility creation and maintenance powered by locally based, farmer-directed, natural processes; and they stand in contrast to externally introduced, purchased stimulants from elsewhere. Those traditional soil fertility management processes are nothing new in the world of agriculture, and the common theme among them is the importance of soil organic matter. In one incarnation or another and across many different cultures, they have supported successful food production since the earliest farmers.

These days we maintain soil fertility and high levels of production at Four Season Farm solely through those age-old, well-tested processes I list in "A Concise Glossary." I refer to these on-farm soil fertility creation and management processes as *information inputs* rather than product inputs. We have determined to be ever more self-reliant and maintain the fertility of our soil without purchased compost or manures. Why should this farm be importing extra organic matter to supplement homemade compost when it could be growing that organic matter right on the farm with green manures? We now shallowly incorporate all crop residues in situ, right after harvest (which means we avoid the work of hauling them to the compost area). Then we plant a grass or legume green manure that is part of a carefully thought-out sequence of year-round, grown-on-site, soil-improving crops keyed to the vegetable crop that will come next in the rotation. We want this farm to be so empowered and self-reliant that it can continue feeding our neighbors in perpetuity no matter what economic forces may affect the stability of the world around it or the availability of purchased inputs. This desire for a secure food supply explains our deep interest in wanting to have a self-fed farm.

CHAPTER TWO

The Start of Purchased-Input Farming

During the early years of European settlement of the North American continent, most European immigrant farmers tended not to worry about maintaining the fertility of their soil, even though this was a concern they had been very aware of in their home countries. Rather, they shamelessly exploited for profit the bounteous natural fertility of the virgin soils they encountered in the New World. Thus, many of them practiced a primitive soil-mining form of agriculture with no thought for the future. And they continued to farm that way until yields fell too low, at which point they would move west to new virgin soil and start over. Once the United States was sufficiently settled, moving on was less of an option; the search for fertilizing inputs began in earnest. As Steven Stoll explains in *Larding the Lean Earth* (2002), "Either farmers could continue to depend on fresh land without a thought of restoration, or they would have to find some source other than the eighty head of cattle necessary to keep a quarter section continuously manured."[1]

The first major fertilizer input collected en masse and sold to farmers, starting around the 1830s, was guano—dried seabird droppings from deposits hundreds of feet deep that were well preserved on numerous rainless islands off the coast of Peru. Guano supplied nitrogen, phosphorus, and trace elements in very available forms. It became such a craze that many countries, including the United States, sent their navies to scour the world's oceans to locate islands with similar seabird deposits so they could claim the deposits on those islands for their own use.[2]

Stoll explains how the guano craze was the beginning of agriculture moving away from the basically self-reliant, farmer-managed natural cycles of the past:

> *If animal manure completed a cycle on the farm, even one complicated by the increasing amount of produce going to market in the nineteenth century, guano represented a breaking and straightening of that line. With guano and all that followed it, the biological foundation of American society became a one-way transfer of material from some point of extraction or production to the farm where it went into the crops, and from there to consumers. . . . With this, more than any other modern addition, cultivation began to change. It began to depend on institutions beyond the furrow—not simply for markets or transportation but for the continuation of the biological process itself.*

In other words, Stoll writes, guano "filled the same niche and arrived in the same way that synthetic chemicals would soon arrive. . . It prepared farmers for what was to come."[3]

And come it did, especially by the mid-1870s when the guano island deposits were exhausted. The chemical fertilizer industry began its growth, and soil fertility concerns began their shift from farm-managed processes to commercially purchased products.

I believe the first product that might be classified as an artificial chemical fertilizer was the superphosphate manufactured by Sir John Bennet Lawes of Rothamsted starting in 1843. It was created to make phosphorus more available to plants by treating bones and other phosphorus sources with sulfuric acid. This process changed the less-soluble compound tricalcium phosphate to the more-available monocalcium phosphate. (Previously, bone residues had been steamed to make their nutrients more accessible to soil microorganisms.)

The other popular purchased chemical fertilizer in the mid-1800s, Chilean nitrate, was mined from the deposits in Chile's northern Atacama region; it became the next guano-like craze. The use of Chilean nitrate is a classic example of the industrial takeover of natural systems that often accompanies a technical shift—the substitution of a new marketable *product* to supply what was previously obtained from the natural world through a biological *process*. Industry thrives on selling products that can be easily packaged and sold. Processes, however, must be performed by people on-site—processes can't be easily packaged and sold. The shift of farming away from the on-farm process of growing leguminous green manures—which take free nitrogen from the air and add it to the soil—in favor of purchasing a nitrogen fertilizer product like Chilean nitrate is a classic example of a product replacing a process. And, in this way, purchased fertilizers eventually took over agriculture.

Inspiration from Edward H. Faulkner

I was inspired to seriously question the ag industry–created mantra of continual dependance on purchased inputs for soil fertility (a chemical farming mindset unwittingly adopted by organic growers that should have been questioned long ago) from

By mid-April, we are already making beds and preparing to plant fields after shallowly tilling in a green manure crop. In the foreground, an overwintered cover crop is greening up.

reading Edward Faulkner's follow-up book to *Plowman's Folly* (1943), which is entitled *A Second Look* (1947). I consider Faulkner to be a very important voice in the philosophy of organic farming. He was a well-educated farmer-scientist who conducted trials on his own farm to prove out his theories.

However, the numerous publications and websites today that celebrate Edward Faulkner as the father of no-till should be seriously embarrassed that they didn't read past the cover of *Plowman's Folly*. Despite the title of the book, Edward Faulkner was not opposed to tillage. He was criticizing only the detrimental effects on soil fertility that resulted when the moldboard plow was used as the tillage instrument, specifically burying the surface organic matter in an airless layer at the bottom of a furrow. In fact, Faulkner enthusiastically celebrated the benefits of shallow *non-inversion* tillage. His radical contention, which he explained more carefully in *A Second Look*, was that shallowly incorporating green manures and crop residues into the top few inches of the soil, either with horsepower or tractor power, using tools such as a disc harrow, a spring-tooth harrow, or a rotary tiller, was the simple solution to maintaining soil fertility. He stressed how easily all farmers could do this on their own farms. And he did not think that farmers who planted green manures needed to supplement their soil with chemical fertilizer inputs. In *A Second Look,* Faulkner wrote:

> *The earth is completely self-sufficient for nourishing the life it develops. . . . The future of the human race depends upon the self-sufficiency of the soil that grows our crops. . . . We may be sure that unless soil really is self-sufficient, its future complete exhaustion is predictable, regardless of future farm practice in the use*

> *of fertilizers. . . . Nature hands to humankind the key to this limitless store of plant food—the many acids that develop in the soil as organic matter decays. Including carbonic acid, we have more than a dozen organic acids that attack the soil mineral crystals, etching away in solution sustenance for plants. Soil yields nutriment in proportion to the quantity of these acids. Further, the amount of these acids depends in turn upon the quantity of decay that occurs. . . . In other words, the more organic matter that has been mixed into the surface of a soil to generate these acids by decay, the more plant nutrient elements the soil will "manufacture" precisely where the plant roots will be hunting for them. . . . Soils cannot be deficient in minerals; they can only be delinquent in giving them up. And this delinquency will begin to diminish immediately the necessary well-distributed organic matter has been supplied in quantity.*[4]

To an agricultural scientist in the 1940s, whose mind was steeped in the conventional scientific thinking of the day, Faulkner's critique of the plow and his dismissal of the need for chemical fertilizers must have sounded unbelievably radical. Add in Faulkner's contention that pest problems are minimized when crops are grown on fertile soils filled with shallowly incorporated organic matter and he sounds even more extreme. (Note that Faulkner's observations about the role of carbonic acids and other organic acids in nutrient cycling have been borne out by much subsequent research; I discuss this in more detail in chapter 3.)

However, in my case, with years of experience seeing firsthand that organic soil management makes my crops very pest resistant, I

Green manures like this combination of bell beans and oats (*left*) contribute significant quantities of biomass belowground as well as above, as this vigorous rye-and-vetch root system (*right*) demonstrates.

had no trouble agreeing with Faulkner's claims. Rereading *A Second Look* was the major inspiration for my conclusion that a self-fed farm, freed from reliance on commercial inputs, might be seen as the ideal for organic farmers to pursue. An ideal that could possibly guarantee the almost perpetual productivity of those farms.

I am not alone in considering this a goal worth pursuing. In his perceptive book *Gardening with Nature* (1954), Leonard Wickenden, a very intelligent and unique voice among the early

US organic enthusiasts (unique because he was trained and worked as a professional chemist and was a former president of the American Chemical Society), dealt with the logical objection to Faulkner—if you are selling produce off the farm and not replacing the nutrients removed by the crops on a daily basis through importing fertilizers, you will exhaust the soil. Well, hardly so, suggests Wickenden, because:

> *If an acre of land contains up to 200,000 lbs. of potash, is it likely that 200 lbs. of potash (which would be all that a generous dressing of pulverized granite would supply) would make much difference? If we do as Mr. Faulkner advocates and incorporate in our soil an abundance of organic matter, perhaps evidence of depletion will not be perceptible for 10,000 years, so the problem can scarcely be considered pressing. . . . Working on wastes from animal and vegetable life, hordes of microorganisms, in numbers beyond the comprehension of the human mind, produce from death and corruption a sweet-smelling soil, richly stocked with compounds to guard the health of vegetation not yet grown and with acids to dissolve minerals and prepare them as food for crops yet to come. . . . Let your aim be to feed your soil—not your plants. The modern method of using the soil as an inert medium for conveying plant food to the crop is grossly unscientific. Feed the soil with organic matter and it will convey well-balanced food to the crops in a steady stream throughout the growing season. There will be no brief stimulation of the plant . . . but a process of day-by-day nourishment which will produce sturdy vigor in the crop.*[5]

Cabbage
artichoke
Carrot

Note that many local soil fertility resources may be available right on the farm itself without our being aware of them. In the case of Four Season Farm, when we hired a large excavator to create a ⅛-acre pond for irrigation 20 years ago, the silt-muck soil that came from the top 3 feet (90 cm) of the semi-swampy area where we were digging had a very low pH but was surprisingly high in phosphorus. The deeper marine-clay layer underneath had an alkaline pH and tested high in both calcium and potassium. Since digging that pond, we have used our manure spreader to distribute reasonable quantities of the stockpiled silt-muck and clay materials from that excavation onto our few acres of sandy, vegetable-producing land. The results have been consistently positive. We have not only been amending topsoil nutrient levels. According to numerous published studies in my files, the intentional addition of clay to sandy soils like ours can modify their productiveness by increasing nutrient-holding capacity.[6]

I celebrate Edward Faulkner's hypothesis of a practically inexhaustible natural soil, enlivened and made fertile solely by the decomposition in the topsoil of green manures and crop residues grown on-site, because it mirrors my hopeful vision of Four Season Farm as a perpetually dependable source (long after I am no longer farming here) of the purest, most nutritious organic produce. Our self-fed farm experiment is well underway in the third decade of the twenty-first century. I hope to be able to let people know, sooner than in 10,000 years, how successfully it is working out.

Early Use of Inputs at Four-Season Farm

Back in 1966, when I first became involved in farming, the nitrogen, phosphorous, and potassium (NPK) chemical thinking about soil fertility had become so thoroughly established that popular organic books and articles were mostly focused on explaining which organically acceptable soil fertility enhancing products could be used instead of the highly advertised NPK chemical products. For example, the organic products they suggested using for nitrogen (N) were soybean meal or dried blood instead of sulfate of ammonia, bone meal or phosphate rock instead of chemical superphosphate for phosphorus (P), and greensand or the product Sul-Po-Mag instead of potassium chloride for potassium (K). During my earliest farming years here in Maine, and in my first book, *The New Organic Grower* (1989), I followed that format of using organic substitutes for standard

(*continued on page 28*)

This scene at Four Season Farm is an example of multifaceted farming in high tunnels, low tunnels, and in the field.

Edward H. Faulkner: 1886–1964

Before I began writing this book, I collected all the background information I could find on Edward H. Faulkner. Given his level of notoriety in the 1940s following the publication of *Plowman's Folly,* there was one surprising item I was unable to locate—a major obituary in the national press. Even though his books were still in print in the 1960s, it seemed as if the counterforces against him had won: His work had been so totally dismissed and forgotten that no one took notice of his death. However, it is not surprising that "conventional" agriculture would have been delighted to see him disappear. *Plowman's Folly* was about far more than his objection to the actions of the plow. He opened the book by expressing a budding interest in other alternative ways of thinking that could be seen to threaten the status quo in many areas of modern life. In just the first dozen pages of *Plowman's Folly,* he not only discredits the plow but additionally "brings virtually all of our soil theories up for critical examination"; condemns our use of excess machinery "to the end of destroying the soil in less time than any other people has been known to do in recorded history"; states that in our focus on adding mineral fertilizers we have ignored "the almost automatic provisions of nature for supplying plants with complete rations in secondhand form" (that is, decomposed plant residues); claims that the values of green manures have been ignored because "plowing down great masses of green manures proved such a colossal boomerang" (the plow was poorly suited for putting under tall rye straw, for example); stresses that "the use of nitrogen—any purchased nitrogen, in fact—is sheer waste of money"; states that if we "recharge the soil surface with materials that will rot," erosion can be prevented because "natural processes will do the rest"; and finally, affirms that present day agriculture causes "both land and people to become poorer and poorer; and people become more and more subject to ailments which we now know are caused by insufficiency of some essentials in their diets." Whew, talk about creating enemies.

Unfortunately for Faulkner's lasting influence, no effort of which I am aware seems to have been made after his death to collect and save his papers and his library. Today's agricultural historians have little idea with whom he communicated and no deep background on what outside influences affected his thinking. In addition, the national environmental organization, Friends of the Land, the leading support structure for most of his ideas at the time, had closed in 1954. Chemical agriculture was on a roll at the time of his death ten years later and seems to have succeeded in making him a nonperson.

chemical fertilizers, and I must admit that it allowed this farm to successfully enjoy exceptional yields of flavorful crops.*

In the first years of Four Season Farm, we needed to quickly build up the poor sandy soil, which we had recently cleared from coniferous spruce and fir forest, so it could grow marketable crops. Thus, we brought in the organic matter to make that possible. It was an enormous amount of work. We had access to a very large horse manure pile on a neighboring farm and I forked many loads of that material day after day into the trailer behind my 1948 Jeep, hauled them 15 miles to my farm, spread it on the fields, tilled it in, and went back for more. I could cut poor quality hay from local abandoned fields and bring that in as a compost ingredient. We added limestone, and in addition we supplemented the soil with phosphate rock. If we had not been in a hurry to put this farm into production, we could have grown deep-rooting sweet clover or forage radish for a number of years to extract adequate phosphorous from the subsoil. After five years we had exhausted the horse manure supply, but with the spoiled hay and crop residues, we kept making compost and had the beginnings of a productive soil. We maintained that soil fertility with green manures and our homemade compost. We also began purchasing additional cow manure compost, but solely from an organic dairy because we were suspicious of the quality of other compost sources.

---

* In subsequent editions of *The New Organic Grower*, I left those recommendations more or less intact (because they worked so well for us years ago), but with increasing suggestions to the effect that "once a vigorous soil biology is established, we are able to maintain fertility with very minimal input systems."

Four Season Farm's portable farm stand and selling at farmers markets are also a part of our farm's story.

It was the expense of that purchased compost ($75 per yard, delivered), and the fuel use for trucking it to our remote location, that first started us thinking about the possibility of growing all our own soil fertility. After a lot of reading, researching, studying, and trying out many potential green manures, the idea of the self-fed farm was born. I have often wondered, given the almost nonexistent soil we had at the start, if we could have created this

A nutritious fall feast can come from the well-managed organic farm or garden, grown without any reliance on purchased outside fertilizers.

farm's fertile soils with the same green manure techniques by which we are now maintaining them. I believe we could have, but back at the start I know I did not have enough experience to pull that off quickly enough to put this farm in business. But it would have saved me the enormous amount of work involved in transporting all that organic matter. (Out of curiosity, I am presently running a trial on two recently cleared acres to determine how long it will take a devoted green manure program to bring that soil's fertility up to our standards.)

From the point of view of reducing energy use, an important consideration in today's world, our present passion for grown-on-site green manures for maintaining soil fertility looks very wise compared with the enormous amount of fossil fuel expended to truck in manures and composts from off site. In addition, green manure crops continue growing while I sleep, and the organic matter and the nutrients that they make available are right there in my soil where I want them.

My present definite preference for maintaining soil fertility with grown-on-site organic matter led me to add a short chapter in the third edition of *The New Organic Grower* (2018) on the benefits of ley farming. I called that brief new chapter "The Self-Fed Farm," and the title stuck as I expanded that line of thinking into this book you now hold in your hands. (If you are wondering what ley farming is, I discuss it in the "Toward Perpetual Soil Fertility" section on page 48.)

I have always been dismayed by the misconception that classical organic farming was focused on "input substitution" rather than on the more substantial and extensive "system realignment," which relieved farmers from dependence on purchased outside fertilizers, and is made possible by employing the traditional proven cultural practices like growing green manures, using cover crops and crop rotations, and so forth. I suspect that this input-substitution critique arose from the assumption that organic *farming* followed the purchased fertilizer input model of organic *gardening*. It traditionally did not.

As mentioned earlier in this chapter, farmers and gardeners in the United States were first exposed to information about the organic method by Rodale's *Organic Gardening* magazine, which presented sound content in many of its articles, but which also included copious advertisements for organic fertilizers. That put the focus on purchased inputs, which explains the attention paid to OMRI—the Organic Materials Review Institute—in the United States. OMRI publishes extensive lists of purchased soil fertility inputs deemed acceptable under the dictates of USDA organic certification. In my opinion, the very existence of OMRI gives the impression that purchased organic inputs are the original foundation of organic farming. They are not. In addition, if I'm not importing materials from outside my farm, I obviously have no need for long lists of which specific inputs are allowable. Organic farming would greatly benefit from supplementing OMRI with an organization I will call OTRI: the Organic Techniques Review Institute. That organization would publish data reemphasizing the importance of green manures, crop rotations, cover crops, growing legumes, and incorporating organic matter with shallow non-inversion tillage. In short, it would return the emphasis of organic farming wisdom to the importance of processes rather than products.

Resisting the Lure of Diluting Organic

All farmers are exposed to constant advertisements in favor of purchased inputs. Only those farmers with a deep understanding of how organic farming really works can see through the hype of that sales pressure and successfully maintain self-reliant farms. The organic movement was created over 100 years ago by ethical farmers who instinctively understood the crucial relation between soil quality and food quality. Unfortunately, the influence of those ethical farmers was quickly marginalized after the organic label became big business and the marketers and merchandisers pushed in and took over. That was the beginning of the end. The merchandisers did not have the same ethics.

One glaring example is the huge wholesaler United Natural Foods Incorporated (UNFI). UNFI employed a lobbyist who wrote a blog that focused on controlling any ethical concerns brought up by farmers or the public. In a segment published in August 2016, the blog writer strongly objected to a lawsuit by the Center for Food Safety. That lawsuit was aimed at preventing contaminated organic matter and concentrated animal feeding operation (CAFO) manure from being allowed in compost spread on organic farms. The blog demeaned the Center for Food Safety's lawsuit as being based on "perfectionist objections by uninformed people who expect obsolete purity in their food." The blog went on to say, "Synthetic materials should be allowed because the very world we live in is contaminated. How perfect can compost be in a polluted world?"

Let me be blunt here; no real organic farmer I have ever known would have written those lines. The blog continually tried to intimidate any farmers attempting to defend old-time organic thinking, by accusing all criticizers of participating in "a circular firing squad"—the lobbyist's pop phrase. In other words, since the merchandisers now controlled organic, and since maximizing the amount of product for sale had become far more important than how it is produced, if you said anything at all

We grow some crops under the cover of greenhouses at Four Season Farm, but always in natural, biologically active soil.

about the diminished quality of the produce, you would be harming organic sales.

Fortunately, there are a number of us old-time organic farmers who maintain those original ethics and still believe in the important relationship between soil quality and food quality. Organic farming began as a statement of faith in the nutritional superiority of food grown on biologically active, fertile soils. The early organic farmers knew that quality food can only result from the quality soil care practiced by the good farmer.

Today's organic farmers did not invent the concept of organic agriculture. It is a gift from a hundred years of development by wise people who farmed before us. We are the beneficiaries of the intuition, experimentation, and dedicated efforts of our predecessors who were concerned with the detrimental effects on food quality caused by industrial methods. They developed the art and science of organic farming because they understood that proper nourishment of human beings can result only when soil is properly nourished. Organic farming is best defined by the benefits of growing crops on a biologically active fertile soil rather than by its rejection of unnecessary chemicals. Crop resistance to pests and diseases is an outcome of farming a soil that fully nourishes the crops.

The importance of fertile soil as the cornerstone of organic farming is threatened these days by the USDA's decision to allow vegetables grown in hydroponic systems to be certified as organic. Just when today's agronomists and nutritionists are finally becoming aware of the crucial influence of soil quality on food quality, the USDA is trying to unilaterally dismiss that connection by removing soil fertility from the National Organic Program definition of organic. The encouragement of "organic" hydroponics is just the latest in a long line of USDA attempts to subvert the nonchemical promise that organic farming has always represented.

I have no argument with hydroponic or aquaponic systems, although I think those who grow that way should proudly advertise that their produce is grown in artificially fertilized water or in fish manure soup. But I am opposed to that produce being labeled organic. If soil-less systems are a valid benefit to the nutrition of the customer and the sustainability of agriculture, then hydroponic operations need to be up front about what they do and work on creating their own sense of integrity and trust rather than trying to illegally piggyback on the organic label.

Let's look at an example of the marketing of another agricultural product that illuminates the inherent wrongness of a hydroponic grower wanting to fraudulently enter the organic market. Let's say a mediocre US winery wants to enter the premium wine market. They decide to use a label that says Château Lafite Rothschild. They blend 80 percent Cabernet Sauvignon with 20 percent Merlot and age the wine in French oak barrels for 18 to 20 months (just like the real Lafite does). Their spin doctors claim that this is desirable because the market wants more Lafite. But, of course, this strategy is illegal, and that wine would be instantly derided and laughed off the market because it was not produced in the authentic manner.

Organic production is equally as meaningful and significant as respecting the terroir of traditional wines. If the fix were not in with the USDA, the idea of certifying soil-less hydroponic strawberries, tomatoes, or lettuce as organically grown would be just as quickly exposed as ridiculous.

I am part of a group of old-time organic farmers called the Real Organic Project who are battling against the USDA for allowing hydroponic produce to be sold as "certified organic."

The enabling legislation of the US organic standards, passed by Congress, states clearly that maintaining and improving the fertility of the soil is a requirement for organic certification. Traditional organic farmers understood that crops grown in water or some inert medium utilizing liquid fertilizers would not be the same as soil-grown crops. Hydroponic food can no more duplicate the nutritional quality of soil-grown food than artificial baby formula can duplicate the nutritional quality of human breast milk.

Organic has been protected by legal definition, but our problem today is that the law is not being enforced. The pressure on the USDA by the hydroponic industry has undermined the organic label in the United States. Allowing hydroponics to be certified as organic is fraud, pure and simple. But there is a serious side effect of that fraud that needs to be pointed out.

The word *organic* is not simply a marketing term. It defines a system of agriculture that acknowledges the preeminence of soil life as the power behind plant growth and quality food production. A wise observer has noted, "Human civilization owes its existence to six inches of soil and the fact that it rains." Only one agricultural science—traditional organic farming—has shown the ability to maintain and enhance the productivity of those 6 inches (15 cm). On a planet with diminishing resources, real organic farming is the only honestly sustainable solution to feeding its people. In contrast, hydroponic growing is an unsustainable technological gimmick completely dependent on inputs and an enormous consumption of energy to support artificial lighting and pumps and filters and controllers and the use of conditioners and additives. Hydroponic is the antithesis of organic.

Real organic farming does not need purchased inputs because it can endlessly renew soil fertility with time-honored practices

This is what the soil on a real organic farm looks like.

that nurture the boundless energy and logic of the earth. Real organic farming is cyclical and can succeed wherever there is soil. Its age-old ecologically sound agricultural practices are accessible at no cost to farmers everywhere because they are based on management decisions, not purchased inputs. The same cannot be said for the technologically fragile, energy-intensive, input-based, factory-food systems that are falsely claiming the organic label.

It is not just the illegal use of the name *organic* that should guide opposition against allowing hydroponics to be certified.

MESCLUN MIX
1/25
DOWN

The greater concern must be to preserve our understanding of the living soil agriculture promoted by classical organic farming so there can be no confusion about which farming techniques can be depended upon to successfully feed future generations in perpetuity. The clear path to a bounteous well-fed future for humankind will remain unrealized if we allow any misunderstandings about—and subversions of—the foundational concepts of real organic farming to go unchallenged.

CHAPTER THREE

The Quest for a Permanent Agriculture

I define and judge the value of my self-fed organic farm on very straightforward grounds: its dependability. However, since Four Season Farm no longer purchases and uses any of the USDA-approved organic soil-fertility-stimulating inputs, considered by many to be indispensable production aids for organic vegetable growing, my choice of the word *dependability* may seem odd. But it is a simple proposition. If I can successfully grow bounteous, clean food harvests year after year without needing to purchase soil fertility inputs from outside my farm, then I alone, not the ag industry, am master of its destiny. My farm's production cannot be interrupted, constrained, or limited by outside market manipulations, supply shortages, price increases, or delivery difficulties.

The self-fed farm concept is what defines the organic standards that I follow. Even if I thought soluble chemical fertilizers would be useful, I cannot produce them from on-farm resources, so they are obviously not part of what I do. What powers the self-fed system is what I can produce within it: organic matter grown on site, along with the plant-nourishing and soil-mellowing humus that results from that organic matter's decomposition by soil organisms. Read as many years far back as you can find in old farming literature, whether from Asian, Egyptian, Babylonian, Greek, Roman, Arabian, or European sources, and you will find that the universal solution to dependable soil fertility, which has been stressed repeatedly over the centuries in every culture, is to maintain soil organic matter. Fortunately, organic matter can be maintained on all farms simply by plant growth in a process that combines energy from the sun, water and minerals from the soil, and CO_2 from the air.

Managing Soil Fertility

The benefits from those age-old natural soil fertility enhancement practices (listed in "A Concise Glossary" on page 10), done within the farm, arise from perceptive management of soil biology, not from purchased inputs. As far as I am aware, those traditional practices, although used by many farmers for generations, had

Vigorous stands of green manures like this young buckwheat at our farm evoke the style of farming practiced by organic pioneers of a hundred years ago.

not been considered a coherent and complete agricultural system prior to about 1900. That thought process, which we now call *organic farming*, began in Europe during the second half of the nineteenth century in counterreaction to the increasing interest in artificial chemical fertilization following the 1840 publication in Germany of Justus von Liebig's book *Organic Chemistry in Its Application to Agriculture and Physiology*.

By the late nineteenth and early twentieth century, a growing number of farmers, who doubted the wisdom and rejected the expense of soluble chemical fertilizers, had begun to establish the foundations of organic farming by rescuing those time-tested soil-fertility-enhancement management practices from the oblivion into which they were being cast by the advocates for the chemical approach. The Liebig followers claimed that any benefits from the natural practices could be replaced by chemical inputs. The organic pioneers had learned from experience that was not so. Because they were good farmers who paid attention to their yields and the health of their crops and livestock, they saw how use of the heavily advertised soluble chemical fertilizers resulted in diminishing the long-term fertility of their soil as well as diminishing the nutritional quality of the resulting crops.

The organic pioneers revised their farming accordingly. Their nonchemical successes demonstrated the superior ability of the age-old practices to create and maintain a level of biologically based natural soil fertility that could grow more nutritious food than the chemically based artificial fertility—and, in addition, do so without the expense of purchased fertilizers. The progressive development of nonchemical "organic farming" as a separate and defined system of food production continued to evolve in European countries because that was where there was an opposing chemical system to react against.

That complicated, industrial-exploitive model of food production—chemically based, pesticide laden, purchased-input dependent, and expensive—had taken over agricultural science almost completely by the 1920s and was being taught by every university and endorsed by national agriculture departments as the only option. But in the 1930s, an increasing number of successful organic farmers began effectively demonstrating a new reality. The organic revolution showed farmers another way.

The creation of organic farming as a new agricultural science was a gift to the world from the combined efforts of many inspired farmers, gardeners, researchers, and advocates for good food. They accomplished a small miracle that offers universal access to food for everyone.

The early organic practitioners brought the science of food production down to earth and back into the hands of the farmers by emphasizing the high quality of crops from healthy soils fed with farm-generated organic matter. They demonstrated how a biologically based agriculture could create and maintain lasting soil fertility by relying on age-old peasant knowledge about the importance of returning organic wastes to the soil. They reestablished the link between healthy fertile soil, healthy plants, and healthy people that had sustained humankind for centuries. Their efforts stressed how food production, when based on biology rather than on chemistry, was in harmony with the natural systems of the planet, was freely accessible to and affordable by farmers everywhere, and was capable of nourishing humankind with exceptional food in perpetuity.

Toward Perpetual Soil Fertility

In the early 1900s as the organic farming system was just beginning to be developed, there was a lot of concern worldwide about the declining state of soil fertility. In 1898, Robert Elliot, a British farmer, had written a slim volume entitled *The Agricultural Changes Required by the Times, and Laying Down Land to Grass*, later retitled *The Clifton Park System of Farming, and Laying Down Land to Grass* (1908), as his contribution to that discussion. Robert Elliot was one of the earliest godfathers of what I call the self-fed-farm concept. For many years Elliot had successfully demonstrated how seemingly perpetual soil fertility could be maintained without purchased inputs. He did this by rotating crops and pastured livestock. For a four-year period, he grazed livestock on deep-rooted grass and legume pasture sod

The offerings at our farm stand are proof of concept for my self-fed farm approach inspired by the writings of innovative farmers such as Robert Elliot and Peter Henderson.

Three feet (90 cm) tall and producing more biomass every day to feed the soil and future vegetable crops.

(following rotational grazing principles). In the subsequent four years, he rotated that same area—after tilling under the sod—to a different use: growing annual crops such as grains, beans, and vegetables. The organic matter added by the roots and tops of the tilled-under pasture plants, plus the minerals their extensive roots extracted from the deeper soil levels, plus the manure deposited by the grazing animals provided ideal soil conditions for the four years of annual harvests (small grains and fodder crops) that followed the tilled sod in rotation, as Elliot explains here:

> *The system is an extremely simple one. It consists in creating, with the agency of large-rooting and deep-rooting plants, a good sod to feed grazing animals, and then relying on that sod, after tilling it under, for the manurial and physical conditions necessary for growing two green and two cereal crops, after which the land is again laid down to grass, and the creation of a good sod again commenced.*[1]

This concept was initially called ley farming or convertible husbandry or alternate husbandry because the fields alternated between increasing the soil organic matter during the pasture years and exploiting that increased organic matter to grow vegetable and grain crops for market during the tilled years. The word *ley* (originally *lea*) is an old Anglo-Saxon word for a temporary pasture that could be converted as needed to tilled crop land. In the United States, this pasture-to-cropland system is referred to as a sod-based rotation.

Peter Henderson, an exceptionally talented US market gardener during the late nineteenth century, had long found that sod was particularly effective at maintaining soil productivity for vegetables. He wrote enthusiastically in an 1884 book that:

> *All experienced growers know that the first year that land is broken up from sod, if proper culture has been given by thorough tillage, the land is in better condition for any crop than land that has been continually cropped without a rest. The market gardeners in the vicinity of New York are now so well convinced of this that, when twenty acres are under cultivation, at least five acres are continually kept in grain, clover, and grass, to be broken up successively, every second or third year, so as to get the land in the condition that nothing else but rotted, pulverized sod will accomplish.*[2]

Writing some 40 years later, John E. Weaver, a professor of plant ecology at the University of Nebraska and a meticulous researcher of root growth in the soil, was equally enthusiastic about the benefits of a grass sod on soil fertility:

> *Grass is a soil builder, a soil renewer, and a soil protector. Covering the land with grass is nature's way of restoring to old, worn-out soils the productivity and good tilth of virgin ones. . . . The humus from the decaying roots helps cement soil particles into aggregates and thus lightens and enriches it. In this way, the mellow texture of the virgin soil is restored, and the accumulation of organic debris, largely from the decayed masses of old roots, adds greatly to its fertility.*[3]

Our growing fields and this small apple orchard are still surrounded by the tall spruce-fir forest from which we cleared this farm.

Robert Elliot had stated in his book that:

> *The economic success of agriculture depends upon the cheapening of production; the cheapest food for stock is grass; the cheapest manure for soil is a turf composed largely of deep-rooting plants; the cheapest, deepest and best tillers, drainers and warmers of the soil are roots.* [4]

Some researchers have criticized ley farming, claiming it resulted in inadequate nutritional value for grazing cattle when pastures were sown with only one grass or one legume, and because of diseases appearing in the stock when chemical fertilizers were overused to stimulate the pasture. Elliot was wise enough to avoid both of those problems by eschewing chemical fertilizers and by sowing very diverse pasture seed mixtures of some 30 or so varieties of legumes, grasses, and edible forbs chosen for deep rooting abilities, "calculated to leave the largest amount of vegetable matter in the soil."[5]

Mighty Mustard green manure (see page 115).

Elliot's biographer, Sir George Stapledon, wrote that Elliot had a:

> *robust aversion to purchasing anything he might be able to produce more cheaply for himself. Elliot therefore set out to devise a system which would be as farm generated as possible in respect to soil fertility and livestock nourishment.*[6]

He went on:

> *The essence of ley farming is to grow crops and grass; and to be at as much pains to use the sod to the best advantage as a manure and the foundation of fertility as to use the grass to the best advantage as a feed. . . . A healthy sod has many of the characteristics of a well-made and well-rotted compost, and management should aim at accentuating these characteristics.*

At Four Season Farm we share Robert Elliot's "robust aversion" to purchasing any fertility we can create for ourselves.

On the US side of the Atlantic, at the turn of the nineteenth century, there was a similar concern to that in England about the general issue of declining soil fertility and a parallel search for a more "permanent" agriculture. Although the USDA Extension Service today has become little more than a sales force for industrial agricultural chemicals, that was not always the case. The director of the Illinois State Agricultural Experiment Station back in 1910, Professor Cyril Hopkins, had his own opinions and did not hesitate to express them.

Cyril Hopkins consistently argued against reliance on purchased chemical fertilizers both in his best-known book, *Soil Fertility and Permanent Agriculture* (first published in 1910), and in numerous extension service pamphlets.[7] Hopkins stressed to farmers how they could save money by creating their own lasting fertility by adding to the soil inexpensive raw materials (such as finely ground phosphate rock) rather than relying on expensive prepared fertilizers such as chemically prepared superphosphate. By "permanent agriculture," Hopkins meant a system under which farmers made the soil better, rather than poorer, through their farming practices. He emphasized that no salesman was going to tell farmers about these ideas because there were no prepared products to sell. In his 1912 University of Illinois circular No. 165 "Shall We Use 'Complete' Commercial Fertilizers in the Corn Belt?," Hopkins wrote:

> *For practically all the normal soils of the United States . . . there are only three constituents that must be supplied in order to adopt systems of farming that, if continued, will increase, or at least permanently maintain, the productive power of the soil. These are limestone, phosphate rock, and organic matter. The limestone must be used to correct acidity where it now exists or where it may develop. The phosphorus is needed solely for its plant-food value. The supply of organic matter must be renewed to provide nitrogen from its decomposition and to make available the potassium and other essential elements contained in the soil in abundance, as well as to liberate phosphorus from the raw material phosphate naturally contained in or added to the soil.*

Turning under the residues of the crop that this greenhouse covered before it was moved.

> *The real question is, shall the farmer pay ten times as much as he ought to pay for food to enrich his soil? Shall he buy nitrogen at 45 to 50 cents a pound when the air above every acre contains 70 million pounds of free nitrogen?*

Stop—read that again. There are 70 million pounds of free nitrogen in the air above every acre of your farm. And that's not all.

> *Shall he buy potassium at 5 to 20 cents a pound and apply 4 pounds per acre when his plowed soil already contains 30,000 pounds of potassium per acre, with still larger quantities in the subsoil? Because his soil*

> *needs phosphorus, shall he employ the fertilizer factory to make it soluble and then buy it at 12 to 30 cents a pound in an acid phosphate or "complete" fertilizer when he can get it for 3 cents a pound in the fine-ground natural rock phosphate, and when, by growing and incorporating plenty of clover (either directly or in manure), he can get nitrogen with profit from the air, liberate potassium from the inexhaustible supply in the soil, and make soluble the phosphorus in the natural rock phosphate?*[8]

Cyril Hopkins was such a well-known and well-regarded professor of agriculture (Hopkins Hall at the University of Illinois is named in his honor) that his name was part of a popular mnemonic to help agriculture students remember the elements needed for plant growth. The mnemonic went "C HOPKNS CaFe is Mighty good," which listed the elements recognized as necessary for plant growth in Hopkins's day.

C-carbon
H-hydrogen
O-oxygen
P-phosphorus
K-potassium
N-nitrogen
S-sulfur

Ca-calcium
Fe-iron
Mg-magnesium

These days most agriculture students are familiar with modern modifications of that mnemonic, which add Mn (manganese), B (boron), Cu (copper), Zn (zinc), Mo (molybdenum), Cl (chlorine), Na (sodium), Ni (nickel), and possibly Si (silicon) to bring it up to date.

Logically the perceptive thinking of both Robert Elliot and Cyril Hopkins eventually found a home with the advocates for organic farming. However, the result of a century of chemical fertilizer propaganda is that hardly anyone today has heard of either Robert Elliot or Cyril Hopkins. The soil fertility facts that they taught, although they were clear to their contemporaries, have been sidelined for so long that they are regarded today as some sort of heresy. (Unfortunately for Cyril Hopkins's long-term influence and agricultural legacy, his strong advocacy against farmer reliance on industrial inputs was cut short following his unfortunate early death in 1919 at the age of 53. He fell victim to complications of malaria during the return voyage after a consulting trip to advise the Greek government on how to rebuild the exhausted soils of Greece.)

Establishing Permanent Fertility at Four Season Farm

Of the three "must have" constituents for "permanent" soil fertility specified by Cyril Hopkins—limestone, phosphate rock, and organic matter—the first two are supplied by finely ground rock powders. As Hopkins notes, the calcium in ground limestone rock corrects the pH of soils that are too acid. Back in 1969 when I first began farming our Maine acreage, our sandy soil, which was cleared from coniferous spruce and fir forest, registered an initial pH of 4.3 (very acidic). Yearly applications of 1 ton of limestone to the acre for the first thee years were necessary for raising the pH into the range most beneficial to plant growth and bacterial life (6.0 to 7.0), especially for the *Rhizobium* bacteria so critical to nitrogen fixation by leguminous crops. These days, we add only small amounts of limestone occasionally to our well-established soil to counteract what slight acid rain effect might be occurring due to atmospheric pollutants that are transported on the winds to the Northeast from Midwestern power plants.

I was aware from the start of our farm endeavor that our land, then newly cleared of conifers, was considered nowhere near as

It might take 100 years for these bits of shell to break down, but they won't require yearly replacement the way commercial fertilizers do. These shells will feed the soil long after my stewardship of Four Season Farm.

promising for farming as land cleared from a deciduous forest like oak or maple trees, which deposit a larger volume of organic matter onto the soil. Thus, I was interested in any possibility of making "permanent" improvements in this land. One fortunate option available locally was a generous daily donation of a thousand or so clam shells (which contain calcium carbonate and micronutrients) from a Maine neighbor who ran a commercial clam shucking business. That continued for several years. I would scatter the shells over whatever section of a field was available and till them in. The rototiller is effective at breaking them into smaller particles. One afternoon as I was spreading shells, the local county agent happened to stop by. "That is ridiculous," he stated. "Those won't break down for a hundred years." I told him that his comment was a perfect metaphor for the difference between the "instantly soluble" thinking of chemical farming as contrasted with the long-term soil fertility thinking behind organic farming. To my mind, I was in the process of intentionally adding pieces of limestone to my soil to duplicate the dependable natural calcium supply of a limestone soil.

The powdered agricultural limestone that many farmers spread on their fields is manufactured according to standards calling for the product to contain specified percentages of ground limestone rock particles able to pass through screens of varying sizes. The finer particles are more quickly effective at raising the pH of the soil. The larger particles are not instantly effective but still constitute a longer-term source of calcium in the soil, the same role I planned for my added clam shells. For a lasting or "permanent" fertility, à la Cyril Hopkins, the necessary mineral (in this case, calcium) merely needs to be in the soil in some form or another. Farmers can then liberate it, in Hopkins's words, "by growing and incorporating plenty of clover" or other organic matter. If we overapply finely ground limestone, it would become available too quickly and raise the pH too high. Instead, we have continued to occasionally add low-solubility clam and oyster shells (or limestone gravel in preparation for a large asparagus planting) to our soil over the years to establish a soil limestone reserve that can be drawn upon as needed by the plants growing in our biologically active soil.

Although limestone was an accepted soil amendment, modern "scientific" agriculture has traditionally discredited Hopkins's second must-have, rock phosphate, as useless because of its low solubility. However, since I am depending on the soil's inherent carbonic acid processes (so celebrated by Faulkner), that "scientific" opinion on the value of rock phosphate doesn't concern me any more than it would have concerned Cyril Hopkins. And I enhance the carbonic acid process by shallowly incorporating the fresh organic matter of green manures and crop residues. Low solubility is a problem only when farmers fail to understand the simple biological techniques (like growing and turning under green manures) they can employ to improve the availability of soil nutrients. In fact, we have taken advantage of that low solubility by occasionally having added an extra dose of "insoluble" rock phosphate over the years to create a phosphorus reserve in our soil. Our calcium and phosphorus reserves are planned as insurance policies for our farm's ability to continue food production during troubled times.

Providing plenty of Cyril Hopkins's third "must-have" ingredient, organic matter from on-farm production, is the focus of this book. Most published agricultural research concurs about the primary importance of organic matter to maintaining soil fertility. There is general agreement that "soil exhaustion" is not the result of exhausting the minerals in the soil but rather a result of not maintaining the soil organic matter. (I expound more on the topic of soil exhaustion in chapter 4.) That truth has been understood for many years by minds that are not captives of the fertilizer mentality. In his introduction to the 1938 USDA Yearbook, *Soils and Men*, Gove Hambidge, the editor, wrote: "It will be evident from the discussion so far that in general nothing is more vital to good soil management than providing for the regular and systematic return of organic matter to the soil."[9] In his chapter in that same volume, William Albrecht, the renowned soil and health expert and one-time chair of the Department of Soils at the University of Missouri, states:

> *Organic matter may well be considered as fuel for bacterial fires in the soil, which operates as a factory producing plant nutrients. The organic matter is turned to carbon dioxide, ash, and other residues. This provides carbonic acid in the soil water, and the solvent effect of this acidified water on calcium, potassium, magnesium, phosphates, and other minerals in rock form is many hundreds of times greater than that of rainwater. . . . Decomposition by micro-organisms within the soil is the reverse of the process represented by plant growth above the soil. . . . Growing plants, using the energy of the sun, synthesize carbon, nitrogen, and all other elements into complex compounds. The energy stored up in these compounds is then used more or less completely by the micro-organisms whose activity within the soil makes nutrients available for a new generation of plants. . . . Organic matter is the source of the power without which the plant-food elements could not be changed to usable form.*[10]

And, finally, author and soil conservationist Louis Bromfield sums up the case in his book *From My Experience* (1955):

> *We should never forget that in a cubic foot of highly productive living, well-balanced soil, every law of the universe is in operation. A living, productive cubic foot of soil is teeming with life, with millions of benevolent bacteria of many kinds, with living fungi and molds, with earthworms and every sort of minute insect life, all in the process of constantly reducing the content of dead organic materials and even minerals into a high state of availability. Within such a cubic foot of soil, the fundamental rule of all life on this planet is constantly in operation . . . the law of birth, growth, death, decay, and rebirth.*[11]

That description explains the beneficial life processes of a fertile soil about as concisely as it can be done. The simple fact is that the key precursors of humus are plant and animal residues and the biological life involved in decomposing them. Both can be grown and maintained on every farm.

CHAPTER FOUR

In Defense of Intelligent Tillage

Anyone suggesting the extensive use of incorporated green manures for creating and maintaining soil fertility on an organic market garden, as I am doing, is obviously going to have to defend tilling the soil. Especially since there have been a lot of recent well-publicized condemnations of organic farming for its use of tillage (many of those condemnations were likely encouraged by the herbicide industry to divert blame from their excessive use of glyphosate). Because I have always found a very positive role for intelligent rotary tillage in creating soil fertility, I was pleased to read the following supportive statement about tillage and green manures in a blog by Andrew McGuire, an agronomist at Washington State University:

> *Green manuring is the lesser-used option in cover cropping. Most cover crops are killed and left on the soil surface, but green manures are tilled into the soil. That is where they have their unique effects. They feed soil microbes and larger organisms and thus change the*

> *community composition to benefit crops that follow. This enlarged soil microbe community then produces more stable aggregates and better soil structure for overall increased soil function. However, for the benefits of green manures to outweigh the tillage required by the practice, a large amount of biomass must be grown.*
>
> *Tillage breaks down aggregates, disturbs soil microbial communities, and quickens breakdown of organic matter. Biomass does the opposite. The more plant biomass incorporated into the soil, the more steps forward the soil takes to overcome the backward steps of tillage.*
>
> *Big biomass also improves green manure's effects on soilborne pests. Both suppression of pests by the chemicals in green manure crops (as with mustard) and general suppression of pests by the feeding of the existing microbe population will be increased by greater amounts of biomass.*[1]

That statement parallels my experience with, and faith in, the benefit of tilling in large biomass green manures. They have been integral to maintaining the biologically active fertile soils on Four Season Farm, and I describe the biomass till method we use in detail in chapter 5 of this book. Over my many years of using rotary tillage, I have seen only positive effects as long as plenty of organic matter is incorporated into the top 4 inches (10 cm) of the soil, either from crop residues and green manures grown in place or from what the Europeans call "cut and carry" green manures, which are grown elsewhere on the farm and brought to the vegetable land.[2]

The rotary tiller is a motorized tool used to prepare soil for planting. It consists of a horizontal, soil-engaging shaft with attached sharp blades that rotate through the soil, cutting and mixing organic residues with the mineral soil particles. The intention is to create ideal physical conditions, provided that the soil moisture and soil temperature are within the desired ranges, for the decomposition of that mixed-in organic matter.

Recent scientific studies reinforce my belief in the benefits of intelligent tillage. Some examples follow. First, a 2007 long-term investigation by John Teasdale of the ARS Sustainable Agriculture Systems Laboratory at Beltsville, Maryland, showed that organic farming can build soil organic matter better than conventional no-till farming because the use of manure and green manures more than offsets losses from tillage. Teasdale stated that the

The cultipacker incorporating newly broadcast green manure seeds; the finished result is the beautifully worked soil shown on page 66.

long-term nature of their study was key to their conclusions because they wouldn't have seen those same results if they had looked after only a few years.

Second is the Soil Health Benchmark Study. This extensive project by the Pasa Sustainable Agriculture organization in partnership with the Cornell Soil Health Laboratory focused on measuring both physical and biological soil health and found that the traditional uses of tillage on an organic farm—weed control and soil preparation—were not detrimental compared to no-till. The study concluded that farms that relied on tillage achieved the same optimal soil health scores as no-till farms.

A third example comes from the Rodale Institute by way of their experience with their initially very strict rules against any tillage for farms that wanted to achieve acceptance under the standards for their Regenerative Organic Certification, a program begun in 2017. It soon became apparent that there were few organic vegetable operations that could meet those standards, and the Rodale Institute had a reality check and changed course. The present standards under the heading "Minimal Soil Disturbance" states: "Soil disturbance shall only occur when necessary to accomplish one or more of these objectives: incorporate crop residues and/or green manures into soil to feed soil microorganisms; control weeds; prepare seed bed and/or planting; break up compacted soil; or develop drainage."[3] That sounds exactly like the logical standards that all of us old-time organic growers, with our concerns about permanent soil fertility, have always followed.

A fourth example can be found in *The Soil and the Microbe* (1931) by Selman Waksman and Robert Starkey:

> *The frequent cultivation of the soil considerably modifies the numbers and activities of soil microbes in*

> *various ways. The processes of plowing, harrowing and cultivating the soil accomplish several distinct purposes: (1) They mix the plant residues with the soil itself, thus leading to better conditions for decomposition of the various residues by microorganisms. (2) They favor aeration of the soil, thus accelerating the exchange of carbon dioxide of the soil air for the oxygen of the atmosphere; the oxygen is required for the various oxidation processes in the soil, especially nitrate formation, the decomposition of organic compounds, and the oxidation of the reduced inorganic substances. (3) Finally soil cultivation brings about a more uniform distribution of the moisture necessary for the activities of the soil microbes.*[4]

That passage brings up another consideration: the age at which a crop is incorporated in the soil. Younger crops with low carbon-to-nitrogen (C:N) ratios will decompose more quickly and provide more nitrogen. Mature green manures with higher C:N ratios will decompose in the soil more slowly because they contain more carbohydrates, fiber, and lignin. For that same reason they will also add more long-lasting organic matter to the soil once they are decomposed.

However, before going further with this discussion, I need to acknowledge that even minimal use of tillage to shallowly incorporate organic matter requires some form of motive power. (On the smaller scale of the home garden, that power can be provided by the gardeners themselves using spades, mattocks, broadforks, and hoes.) A logical solution on a determinedly self-fed farm, which keeps this on-farm soil fertility system powered by on-farm resources, would be to use animal traction to pull a disc harrow or

Transplanting onion seedlings into newly shaped beds marked with a grid.

a non-PTO (power take-off) system, such as a spring-tooth harrow, like the one designed in 1949 by Massey Ferguson following Edward Faulkner's concepts and known popularly as the Ferguson tiller. Both harrows do a nice job of shallow incorporation.

Employing a PTO rototiller powered by an internal combustion engine, as I have traditionally done, means burning fuel. To

The tilther, powered by an electric drill.

avoid relying on fossil fuel (a very desirable goal) we presently run our small diesel tractor on what we call natural diesel, which is a hydrocarbon diesel fuel produced by hydroprocessing agriculturally produced fats, vegetable oils, and waste cooking oils. Another option might be to install solar panels on the barn roof and purchase an electric tractor instead of a diesel. We are continually exploring our tilling options so as to keep our farm productive and self-reliant, no matter what the future may offer.

I have used a rototiller for soil preparation ever since I began farming. Back in 1966 I started with a 4 HP Troy-Bilt rear-tine model as my entry level tiller. (I could even till old sod with that little machine if I made a few passes at right angles.)

I stepped up to a 10 HP BCS once I was earning enough money to afford it—though I never liked its excessively fast tine speed. For Four Season Farm's 3 acres (12,140 sq m) of vegetable production, I purchased a small John Deere 870 tractor with a 48-inch (1.2 m) PTO tiller. I have learned that the tiller manufacturer Maschio Gaspardo makes a tractor-scale tiller with a four-speed gearbox, which allows the farmer to adjust tine speed, and a rear cage roller to give precision depth control for shallow tilling, but I have not yet been able to acquire one at the time of this writing.

I consider the rototiller to be a perfect tool for shallowly incorporating green manures, weeds, well-rotted animal manure, and compost into the soil, and this is how I have used it in the past. It is also ideal for mixing finely ground rock powders (such as limestone, phosphate rock, and basalt dust) throughout the soil. The mixing process is important because it brings those low-solubility sources of major and minor nutrients into extensive contact with all the soil acids and microbiological processes that are involved in increasing the availability of the minerals they contain for uptake by plant roots. And, in addition, a rototiller does the precise type of shallow and thorough incorporation of green manures and crop residues that Edward Faulkner always recommended. In fact, even though they were uncommon in his day, Faulkner was aware of rototillers and made positive comments about them in both *A Second Look* (1947) and his later book *Soil Development* (1952). For home gardeners he recommended the manufacture of small, inexpensive gasoline-driven machines "that would effectively stir into the soil all surface trash."[5]

Many new enthusiastic young organic growers entered market gardening in the 1960s (with no previous experience, as I did). I have often speculated that one key reason why we were so successful was that we used rototillers from the start. They were the easiest and most effective beginning-level machine for soil preparation. None of us could have afforded tractors with plows in our early years, and we were not farming on that scale anyway. I doubt if many of us had read Faulkner's books, and thus the benefits of his highly recommended shallow incorporation of organic matter into the surface of the soil became our preferred preparation method by pure chance, but it obviously contributed to our unparalleled success.

Soil Exhaustion

I have no arguments with basic soil science. I want to make clear that I am not trying to contest or disprove the prevailing "bank account" concept of soil minerals (if you withdraw, you must deposit) that everyone seems to accept. Rather, I want to point out that soil mineral exhaustion is a long, slow process that takes place over many years and must be understood as such. The authorities I cite in this book would agree with University of Minnesota professor Harry Snyder's comment in the 1896 Yearbook of Agriculture: "The most important difference, physical or chemical, between the composition of old, worn soils and new soils of the same character is in the amount of humus which is present."[6] They would also concur with the statement from the University of Vermont in 1908: "Soils are more apt to be worn out because of physical ill condition than because of plant food

Building healthy soil is a worthy project that farmers can accomplish with good management practices. Soil exhaustion is only an issue when soil organic matter isn't maintained.

shortage; a misfortune due mainly to ill-usage. That ill-usage arises from not maintaining the soil organic matter."[7]

This book focuses on techniques that will fully prevent such ill-usage. The way in which one quantifies the supply of minerals in the soil must be considered, however. The average soil depth referenced in most claims about the quantity of minerals available in a natural fertile soil is only about 6 inches (15 cm). These 6 inches, referred to as an acre furrow slice, are the *top* 6 inches of the acre in question. However, if the roots of crop plants reach a depth of 36 inches (90 cm)—which some do—those roots can explore and exploit the mineral nutrients in a volume of soil six times as large as the top 6 inches. If green manure roots reach depths of 60 inches (1.5 m) or more—as some do—there exists a volume ten times as large for roots to explore. The green manure varieties used in our system are selected for their deep rooting ability. By making that choice, we can logically conclude that we are slowing down the time it takes to exhaust the minerals of a soil. And, as Leonard Wickenden noted in *Gardening with Nature*, exhausting that mineral reservoir will take many years, so "the problem can scarcely be considered pressing."

As discussed earlier, I have made provision over time for Four Season Farm's long-term productivity by augmenting its soil with quantities of finely ground rock particles that contain supplies of the minerals that Cyril Hopkins advocated but which are not naturally abundant in the farm's soil. Since most soils are principally composed of particles of ground rock, that practice duplicates the form in which those minerals might naturally exist in the soil. In addition to limestone and phosphate rock, another type of ground rock we have added to the soil over the years as a broad-spectrum mineral reserve is basalt rock dust, a by-product of the production of trap rock that has long been a popular soil amendment on organic farms in Europe. It is usually available for free at basalt quarries as "trap float," which is a very finely ground waste of the grinding process. The basalts are well-balanced rocks that supply a wide range of both major and minor soil nutrients. Basalts weather more easily than granites because they contain less silica and more calcium and magnesium. Soils derived from basalts are rich in clay and iron oxide and are usually very fertile. Basalt rock powder adds to the soil a material that approximates the composition of highly fertile, unweathered young soils. ("Young" soils consist of primary minerals that provide abundant amounts of the essential plant nutrients because weatherable surfaces are still present.)

Back in 1948, W. D. Keller (author of *The Common Rocks and Minerals of Missouri*), writing in *The Scientific Monthly*, considered powdered basalt to be a long-lasting, well-balanced soil builder and a complete plant food. Keller summed up his case:

> *A thorough addition of a balanced rock should last several times as long as an equal application of quickly soluble fertilizer and at no time would there be any effect of overdosage. A rock supplement lasting ten years would be cheap even if it cost three times as much as one applied every year.*[8]

A recent review of the whole subject of ground rock fertilizers, written by three German researchers and entitled "Remineralizing Soils?," also comes to similar positive conclusions.[9] Basalt rock dust is the third slowly available natural mineral source that we have amended our soil with over the years in our quest for a long-lasting, inexpensive, almost-permanent soil fertility that doesn't have to be reapplied annually.

CHAPTER FIVE

Putting Green Manures at the Heart of the System

The first large biomass green manure system we explored at Four Season Farm as we began working to establish our self-fed-farm system was a combination of winter rye and hairy vetch. I call this rye-vetch green manure program *biomass till*. It has been the most important factor in creating and maintaining a very productive farm without outside inputs.

We currently plant this mixture on approximately a third of our 3 acres (12,140 sq m) of vegetable land every fall in mid-October following the autumn harvest of storage crops like carrots, beets, celeriac, cabbage, turnips, and potatoes. In the past, it took until mid-November to store the last of those crops because we had to wait for the temperature to drop down low enough in our traditional root cellar, which is cooled through passive air exchange during cold nights. That late harvest and storage schedule unfortunately meant we could not seed the rye-and-vetch green manure mix on time, as October 15 is the latest dependable

Installing this CoolBot in our root cellar expedited the process of cooling the cellar for winter vegetable storage.

sowing date at Four Season Farm, so we added a CoolBot unit to our root cellar.[1] Now the temperature in our storage area cools down to 35 to 40°F, with all the crops inside by mid-October. Once colder fall temperatures arrive in November, we turn off the CoolBot in favor of the economical cool-air-in, warm-air-out passive air exchange to maintain the ideal cellar temperature over the winter.

With the harvest stowed in the root cellar by mid-October, we can sow our rye-vetch mix on time.

Our October 15 (or earlier) rye and vetch sowing remains in place through the winter, the following growing season, and a second winter. We sow these green manures by broadcasting the seed, walking the fields with a chest seeder just after harvesting the storage crops. We then use a cultipacker to press the seeds into the soil. Our fall rains are dependable, but we could irrigate if necessary. The hardy rye and vetch plants germinate in the fall

Farmer Caleb Hawkins backs the flail mower through a lush stand of buckwheat to shred it down to soil level.

and survive the winter. After they begin vigorous new growth in the spring, we let them grow on to maturity. Our summer farm stand customers are fascinated by the tall, tan rye straw bearing full heads of grain in some of our fields. Few small grain crops are grown to maturity around here these days, and the fields resemble a Currier & Ives farm scene lithograph from a different era.

On or about August 10, it's time to mow the green manure. I watch to see that the mature rye plants are over 6 feet (1.8 m) tall and well headed out and that the vetch vines with full seed pods have climbed the rye straw. Then, on a very dry day, I back my tractor through the rye and vetch with the rear cover on the PTO

(*continued on page 88*)

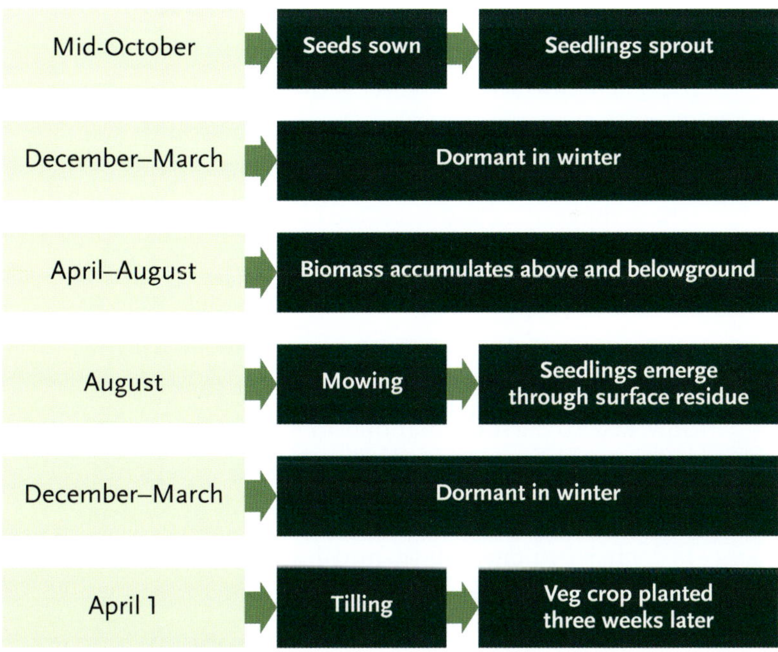

From first planting to final tilling, here is the 19-month timeline for our rye-vetch green manure combination.

Snow Seeding Can Save the Day

Before we upgraded our root cellar, there were a few years where our too-late-planted rye and vetch did not germinate and grow at an acceptable level. We learned to solve that and salvage the green manure value of those fields by "snow-seeding" in winter with Cyril Hopkins's favorite leguminous green manure: sweet clover. Snow seeding and frost seeding are dependable New England techniques. It may seem illogical to be out in February or March broadcasting sweet clover (and other clover seeds—forage chicory also works) with a chest seeder on to frozen or snow-covered fields, but sowing during the winter is the best way to achieve better germination with these crops. As soon as the snow melts or the soil thaws, the seeds fall into cracks or crevices in the soil surface where the moist conditions are ideal for early spring germination. Snow seeding is also the easiest way to tell if you have achieved thorough coverage because the seeds stand out on the snow surface. In fact, if you need practice determining at what pace to walk and how fast to turn the handle on your chest seeder, snow seeding is a great way to gain that experience.

A winter sowing of sweet clover will grow over 18 inches (46 cm) tall by mid-July, when it can be tilled in for a fall brassica crop. With its powerful roots (more than 4 feet [1.2 m] deep at only four months old) and active nitrogen fixation ability, sweet clover is a serious soil builder, especially when grown with a grain crop for extra organic matter. We continue to keep sweet clover seeds on hand for snow seeding in case we need to bail out any other fall-sown green manure germination difficulties. When sowing sweet clover (*Melilotus officinalis*), use an inoculant specific for the species or purchase pre-inoculated seed.

flail mower raised and the depth roller removed to shred all the standing straw right down to soil level. The result is a two-inch-deep (5 cm) mulch of chopped rye and vetch straw covering an enormous number of rye and vetch seeds. The next rainfall stimulates all the seeds to germinate (or we irrigate if conditions stay dry). The seedlings grow up through the straw, and the area soon looks like a golf course fairway.

This beautiful new growth of rye and vetch begins the second winter of root-mass accumulation. Some authorities suggest that root growth is more valuable than top growth for adding long lasting organic matter to the soil. I leave the self-resown rye and vetch until the first of April the following year—nineteen months in total from the original sowing date. By then, enough new, nitrogenous green growth of rye leaves and young vetch plants has formed to aid the final decomposition of the straw, and then I till it all in. I have noticed that young rye plants incorporated at that stage of growth do not become weeds, as they sometimes can, probably because they have been stressed by such close spacing—about twenty times the standard seeding rate. After allowing three weeks for the residues to decompose, the area is ready to start growing that year's vegetable crops. Two winters and one summer of rye and vetch creates big biomass, and the resulting soil fertility improvement is impressive. This process then moves to a different third of our vegetable land each year, following the crop rotation plan.

Other Useful Combinations

The green manure treatment of the fields on the rest of our vegetable land is the other part of this self-fed farm project. As soon as any vegetable crops have been harvested, the residues in those fields are tilled in and then they are sown with green manure crops, some of which will winter-kill in our climate. Examples are bell beans and annual alfalfa, which in addition to being legumes also have deep roots to bring up nutrients from the subsoil; buckwheat, which effectively smothers weeds because of its fast growth and large leaf area; forage radish, which has very deep roots that effectively scavenge minerals from the subsoil; or a

Newly germinated undersown clover filling spaces between rows of cabbage will carpet the ground by harvesttime. After cutting the heads, we will chop the crop residues and incorporate them into the soil along with the clover.

Young arugula undersown
beneath pepper plants.

Spring-sown peas and oats on the right; fall-sown rye and vetch on the left.

combination of peas and oats, which are a great grass-legume winter soil cover. We also use deep-rooted forage chicory if we want a green manure of that type that will survive the winter. For our summer transplanted brassica crops that are harvested from the field late into the fall (like broccoli, cauliflower, kale, and brussels sprouts, which we want to leave productive even later than our storage crops), we scatter legume green manure seeds by hand during the summer and rake them into the soil beneath those standing crops three weeks after we have transplanted them. This practice is known as undersowing.

The dead top growth of the winter-killed green manures provides excellent winter protection for the soil and can be raked aside in spring to be used as mulch or taken to the compost heap, or it can be tilled in before seeding or transplanting the spring crops. The fertility boost from the winter-killed legume green manures gets the early crops off to an excellent start. In the early fall, the year before an area is to be used for growing root-cellar storage crops, we sow hardy leguminous crops such as red clover and winter vetch that survive the winter. They will be tilled-in three weeks before those cellar storage crops need to be direct sown or transplanted. After harvest, those areas will be sown to winter rye and hairy vetch.

The choice of one green manure variety over another depends on the time of year the green manure crop is to be sown (usually between early April and mid-September), expected weed pressure (a reason why we might sow buckwheat), and which vegetable crop will appear next in the rotation on that area. For the summer crops that leave a good bit of post-harvest biomass, such as broccoli, cabbage, and sweet corn, I chop their residues with the tractor flail mower and shallowly incorporate them with the tiller before seeding the green manures. I also use the flail

Squash and clover can share space. Sweet clover is germinating between rows of young winter squash plants.

mower to shred the stems of the very late fall-harvested brassicas once we finish harvesting at the end of November, to get them started on the road to decomposition.

Growing Vegetables and Soil at the Same Time

Since the fertile soil that we are striving to create only earns money for the farm when it's used to grow vegetables for sale, the planting of green manures (with a goal of growing fertile soil) has been criticized for occupying land that could be producing food crops. We have looked at that situation closely from both the perspective of the crops and the perspective of the green manures.

The squash vines ramble over the sweet clover, which eventually pokes up through spaces between squash leaves.

From the Vegetables' Perspective

First, we considered how to keep the green manures from getting in the way of the vegetable crops. One option is to grow a productive green manure like red clover on a field that is not part of the current vegetable rotation and, after mowing and transport, till the green top growth into a vegetable field. After a three-week interval, during which the soil organisms successfully decompose the added organic matter, that field can be planted with vegetables. Obviously, the field won't receive the tremendous benefit from tilling in the root growth of the green manure, but the field won't be tied up for the additional six weeks or more that's needed to produce that organic root matter in situ.

Another option is to focus on green manures that can productively occupy the soil during the late fall and winter months when the vegetable crops either can't be grown or have already

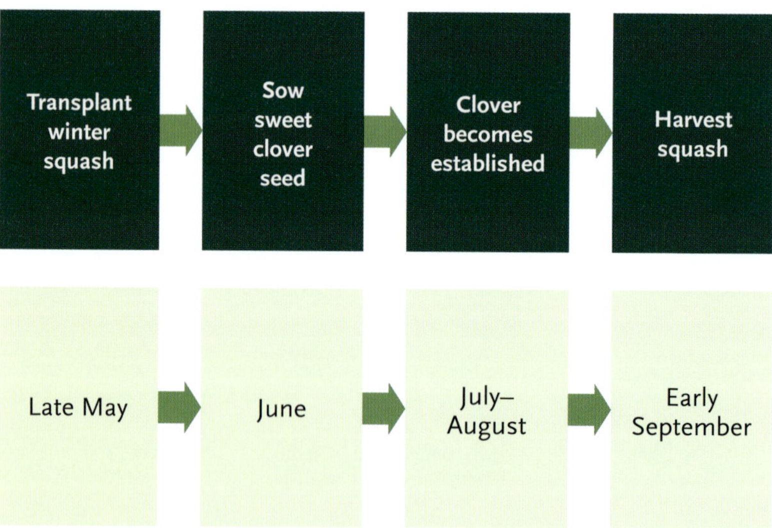

Timing is important when undersowing a green manure beneath a vegetable crop.

been harvested. That period, during which the growing conditions are either too cold or the sun is too low for vegetable crop production, is when our overwintered and winter-killed green manures have been such a success. The one green manure that we allow to occupy soil space over the summer months, our 19-month large-biomass rye–vetch combination, is tolerated because of its enormous contribution of homegrown organic matter from the tall rye and vetch straw. And because we allow it to naturally reseed in the field, it provides a second winter of root growth and soil protection without our having to buy and plant more seeds.

Another tried-and-true solution to keeping the green manures from getting in the way of the crops is to sow green manure seeds under the standing crops, either sprinkling the seeds by hand and raking them in, or using a small multiple-row seeder (such as a Pinpoint Seeder from the online store Johnny's Selected Seeds) or my homemade hand-pushed seed drill. (See "Can Green Manures Outcompete Weeds?" on page 138 for a description of this implement.) The key to this technique of undersowing is timing. Weed competition studies have noted that many direct-sown or transplanted vegetable crops can be kept weed-free by cultivating the soil around them for the first three to four weeks after they are established. Crop plants can tolerate low-growing weeds after that period without yield decline. So you can think of the green manure seeds that you might choose to undersow in that clean cultivated soil around your crops as deliberately planted, desirable weeds. However, don't delay. You want the deliberate weeds to get successfully established and growing well in the understory before the growth and maturity of the crop can create too much shade and competition to overwhelm them. As I said, timing is the key. When undersowing is done correctly, the result is enormously satisfying.

From the Green Manure's Perspective

Next, looking at the situation from the perspective of the green manures, we considered how to keep the crops from getting in the way of the increasingly important soil fertility contribution from the green manures. If planning to establish a productive green manure, it is vital to harvest the previous vegetable crop in a timely way. Some fall-sown green manure varieties need more

This pea-oat green manure crop has been left growing too long. Those pretty flowers could become a pesky weed problem next spring.

weeks to get established than others. Thus, planting the green manure crop soon enough to make good growth before winter is a crucial consideration. Washington State University conducted a planting-date study with mustard green manure in 2002. They documented the progressive decline in dry matter yield of the mustard green manure from a total of 7,458 pounds (3,382 kg) when the mustard was planted on August 13 to a measly 1,829 pounds (830 kg) from a September 3 planting on down to a negligible dry matter yield from a September 10 sowing. The dry matter yield dropped about 30 percent with each additional week's delay in planting after August 13. That data helps make a case for farmers to keep notes of the performance on their farms of each green manure variety they intend to use and to plan the timing of the crop rotation program on each of their fields to incorporate that knowledge.

Of course, it is always possible to plant the green manures too early. For example, a combination of peas and oats is a very popular choice on Four Season Farm because we get the benefits of quick germination from the large grain and legume seeds plus excellent winter soil protection. We need to be careful, though, not to plant this green manure duo so early that the peas and oats mature sufficiently to make viable seed before winter temperatures kill the plants. We want as much bulk as possible for organic matter and winter soil protection but not so much as to risk ending up with a pea and oat weed problem the following spring.

That same consideration exists with a late-summer buckwheat green manure sowing. It is tempting to leave the buckwheat to freeze and become a thick winter soil protecting layer. But green manure enthusiasts are always warning growers to at least mow the buckwheat before it can set viable seed, which could become a troublesome weed the following year.

What Inputs Are Acceptable?

As a specific component of his must-have list, Cyril Hopkins mentioned "incorporating plenty of clover (either directly or in manure)" as a guarantor of soil fertility.[2] Clover has long served as both a green manure crop and as a popular ingredient in grass and legume pastures and hayfields for livestock. Livestock manure that is nourished from the grassland sections of a farm is always considered an important on-farm input to help maintain fertility on tilled land as well as pastureland. It has been celebrated as such since the dawn of agriculture. However, the long-term reputation of livestock manure as an instant soil-fertility-enhancing input has tended to inhibit equal consideration for the soil fertility boost gained from growing green manures or from periodically tilling under those livestock pastures and hay fields as the ley farming systems do. That overfocus on manure as an organic fertilizer was evident from statements made by conventional soil scientists in the early years of the organic farming movement. Most agricultural scientists at that time were totally captivated by their belief in the indispensability of chemical inputs for soil fertility, and they were eager to disparage the practitioners of organic farming. A number of those scientists were quoted displaying their embarrassing ignorance of all the other options that farmers have employed over the centuries to maintain soil fertility by stating bluntly that organic agriculture could never feed the world because there was not enough manure available.

When organic farms purchase manure or compost for soil fertility, those inputs are often referred to as coming from "ghost acres"—land not part of the farm. The designation of ghost acres is useful to point out if one is comparing yield per acre (4,050 sq m) between contrasting systems, but I mention it here because on a mixed farm—both crops and livestock—using the manure from ruminant livestock fed from the farm itself as an input for the tilled land does not disqualify it as a self-fed farm because no ghost acres are involved. The most complete farms are mixed farms.

Chicken Manure

Four Season Farm used to host a few hundred laying hens grazing on a grass-legume pasture in the rotation, and I was pleased with the performance of the crops that followed that pasture section. The dozens of eggs from the pastured hens that we sold every week made sense as a supplementary product on our small acreage. Laying hens on range were the ideal livestock for a small vegetable farm like ours. Since we were already using mobile greenhouses for vegetables and knew how to manage them efficiently, we housed the hens in a mobile greenhouse equipped with laying boxes and roosting bars. The greenhouse moved across land that we had sown to a pasture mix. We could expose a new area of pasture for the chickens to graze daily by moving the house forward down the pasture. In winter we would plow the snow off that area before moving the house. The chickens did not mind the frozen soil.

We kept the hens for only one year of laying to optimize egg quality, and we fed them purely organic feed. At the end of that year, when they started to molt, we sold them to local chefs who wanted to make organic chicken stock. That well-manured pasture then recycled back into our regular rotation, and the next year a new young flock protected by another mobile greenhouse began moving down a new seeding of grass-legume pasture.

Although our eggs were a popular item, two aspects of the enterprise worried us. One was financial: The cost of having certified-organic feed delivered to our remote location was high.

The other was ethical. After reading a *Washington Post* exposé by Peter Whoriskey of the USDA's intentionally lax inspection of supposedly "organic" grain shipments imported from eastern Europe, we came to doubt whether the feed we were purchasing was honestly organic. Together, these issues caused us to reconsider the benefits of that operation, and we have discontinued our laying hens. As Whoriskey's article pointed out, "gauging the extent of fraud in imported organics is difficult because there is little incentive for organic companies to advertise their suspicions about suppliers."[3] I suspect the same might be said of suspicions about the quality of organic inputs.

Compost

I must admit that I have always found our own homemade plant-based compost gives superior results over the cow manure compost that we used to purchase from a local organic dairy. I suspect that the difference arises from the wood products—shavings and sawdust—that are traditionally added to manure to adjust the carbon-to-nitrogen ratio for successful composting. The effect of wood products in the soil has never seemed to be positive in my experience. I once had access to a supply of composted horse manure from horses bedded on straw (to absorb the urine), and I considered that to be a wonderful soil improver. The use of straw as the carbon source in composting seemed to make all the difference. At present, I consider our own homemade compost without any manure added to be pretty darn perfect because we pay a lot of attention to making it well. The fact that we could never seem to make enough of it is part of what led me into my passion for green manures as an equally effective solution compared to animal manures, and green manures require a lot less work.

Straw bale–walled compost bins at Four Season Farm.

The Healthy Plant Theory

The most exciting feature of organic farming is the enlightened understanding one has of the relationship between plants and pests. Insects and diseases are not seen as *enemies*. Rather, they are seen as *indicators*—indicators that the growing conditions are inadequate for the physiological needs of the plants. Pests are bringing a message that the plants are under stress. When we use pesticides we are basically shooting the messenger and ignoring the message. This view of the plant–pest relationship is known as the healthy plant theory or the plant-stress hypothesis. The fact is we see the truth of it on our farm every day. When we have created growing conditions that meet the needs of the crop, pests and diseases become innocuous. Numerous scientific studies support our practical experience.

Minerals

My interest in having a self-fed farm is practical, not irrational. It stems from my desire for perpetual production of exceptional quality food that can't be sidetracked by a sudden unavailability of inputs. Obviously, if there is some element that I realize needs to be added to my soil to grow more nutritious vegetables, of course I will add it.

The one purchased product that we still import and add to our soil, for the crops that need it, is boron. The soil on this farm is designated as a sandy acid podzol on our original soil map. According to data I have read on boron availability, a sandy acid podzol that has been limed to correct the acidity and that has high organic matter levels thanks to additional soil improvement is the most difficult situation for boron availability. Unfortunately, that describes this farm. Even where we have spread seaweed collected from the shore after storms, which we were told would help, we have still seen boron deficiency symptoms in crops of beets, celery, and broccoli. So, to solve that problem, we purchase a product called Granubor (from the company U.S. Borax), which we sprinkle on the soil at the recommended rates. Granubor comes from one of the world's cleanest borate sources. Every batch is made in the USA and is traceable.

Seeds

The seeds for the green manures we are growing can't be ignored here as inputs. At the start of my green manure efforts I purchased all of them. Now, because of an interest in determining just how self-contained this system can be, I have begun experimenting with simple methods for saving my own seeds. I have set aside a small field dedicated to the seed-saving project. Every October, on the day I sow the rye-vetch mixture we use as our major large biomass contributor on our just-harvested storage-crop fields, I also sow the same mixture on a measured section of that seed-saving field, large enough to yield the necessary seed crop. The following summer, when the seeds are mature, I choose a dry day and mow that measured section with my scythe and cradle, the same technique our ancestors used to harvest their grain crops for centuries past. After mowing, I move those sheaves of tall rye and vetch straw on to a large tarp and use a flail to thresh out the mature seeds—another centuries-old technique. I then take the straw to the compost area. As for the mass of mature seeds, I store them in rodent-proof containers in anticipation of a windy day when the brisk air movement will help me winnow the husks and any other chaff from the grains—yet a third old-time technique. At that point, if I have calculated correctly, I will have the adequate quantity of seeds necessary to plant another year of rye-vetch green manure.

We also grow and save the seeds of the other green manure crops upon which we rely as an indispensable part of our fertility maintenance program—bell beans, chicory, forage radish, and sweet clover. They are grown in the seed-saving field and harvested using those same centuries-old techniques. We were not successful saving the pea and oat seeds (our most commonly planted, winter-killed green manure) with our old-time techniques when we grew the two crops together in the way we use them as a winter-killed green manure. We now grow them individually for the purpose of harvesting seeds, and it is a lot easier to capture and save those seeds. We have not yet figured out how to effectively harvest our own buckwheat seeds, because the seeds fall off the plant when we mow the buckwheat with the scythe, so we continue to purchase them.

CHAPTER SIX

Individual Green Manures

Four Season Farm is located on a peninsula about two-thirds of the way up the Maine coast. When we began here in 1969 we experienced a zone 5 winter climate. Now, thanks to climate change it has been averaging around a Zone 6.5 winter climate recently (about 15°F warmer). Because of this change, I am reconsidering whether some of our previously dependable winter-killed varieties may start surviving the winter. Where the winter-killing feature is very important to how and where we use those varieties, I have begun investigating even more cold-sensitive substitutes.

We have an excellent irrigation system on this farm. The water comes from two constructed ponds plus a deep well that can produce 100 gallons per minute. Irrigation is a crucial insurance policy when relying on green manures for fertility. Dry late summer or early fall weather can seriously inhibit successful establishment of broadcast-sown green manures. Drilling the seeds is always the most successful option for sowing, but it requires tractor-scale equipment such as a seed drill, which may be hard to find locally at an affordable price.

Reliable Green Manures at Four Season Farm

The green manure varieties that have worked well for us include a cereal rye–hairy vetch combination, a duo of field peas–common oats, bell beans, buckwheat, forage radish, Mighty Mustard, red clover, and forage chicory.

Cereal Rye and Hairy Vetch

Cereal rye (*Secale cereale* L.) is also called winter rye because of its cold hardiness. Planted with hairy vetch (*Vicia villosa*), the combination serves as our overwintering large-biomass mixture. This pairing has become the major biomass supplier for our self-fed farm. Both have been highly recommended as cover crops for Northeast farms by USDA sources for many years because they are similarly cold tolerant. The two varieties can be sown late in the season; they will germinate and grow at cold temperatures.

Years ago, when we used winter rye by itself (sown in fall and turned under in early spring), we found it difficult to control its vigorous regrowth with the primitive tillage equipment we were using at the time. As described in chapter 5, we now sow this mixture in fall, mow it in early August the following year, leave all the seeds to resprout, and then turn it under the following April (19 months in total). As a legume with excellent nitrogen fixation capability, vetch is an ideal companion for the grain rye. (If you haven't previously grown vetch, use a field pea and vetch inoculant at planting.) They are also well suited to one another because vetch is a vining plant that benefits from being able to climb the strong rye straw for support. During the summer, the vetch vines (covered with seed pods) grow as tall as the rye.

It is important to find the correct ratio between the amount of rye and amount of vetch when sowing. Most mixes use about 80 percent rye and 20 percent vetch. A number of years ago, wondering if we ought to be looking for more nitrogen, I began trials to see if there was a better ratio than that. However, once the ratio reached 30 percent vetch in the mix, the vetch took over and its vines overwhelmed and pulled down the rye straw, creating a total mess. Accepting an "if it ain't broke, don't fix it" lesson from that experience, I returned to the more conventional 8:2 ratio.

Field Peas and Common Oats

Our other reliable green manure mix is a combination of field peas (*Pisum sativum* L.) and oats (*Avena sativa*), sown after August 15 and in the first part of September when so much land

Forage chicory (see page 116).

Field peas in flower.

opens up following the harvest of vegetable crops. This is a vigorous growing mix that serves two purposes: One, it provides a bulk of both grass and legume residues to protect the soil over winter and take up any available soil nutrients before they can leach; and two, it winter-kills by January. The remains can be either tilled in to further benefit the soil, or raked aside so no-till seeding of a spring crop can take place at the earliest possible moment. Because avoiding unnecessary tillage seems like a wise choice, we prefer the second option. However, if we plan to transplant hardy early spring seedlings, we leave the winter-killed residue as a mulch and transplant through it at the appropriate time.

We also plant field peas and oats in early spring on the empty site of one of our moveable greenhouses. They are tilled in four weeks before the August 15 sowing of our famous cold weather–sweetened "candy carrots" (these young, tender, late-sown carrots are protected by a movable greenhouse after November 1 plus an inner layer of fabric after December 1). These carrots, harvested fresh daily from the cold soil all winter, are the sweetest of any we grow. It is our most popular winter crop. The pea–oat green manure has proven to be the best green manure to prepare the soil for this crop of carrots.

Bell Beans

Bell beans (*Vicia faba*) are a small fava bean with seeds the size of a small lima. We sow them in late summer. They are most successfully sown in rows 4 inches (5 cm) apart with an EarthWay seeder (using the lima bean plate) at the same depth as peas. We also use a pea and vetch inoculant. The large seeds allow these legumes to grow rapidly after germination. Research has shown that some fava beans can fix up to 100 pounds per acre (45 kg per 4,050 sq m) of nitrogen in six weeks. Before drilling the favas, we broadcast oats lightly on the bed. The two grow happily together. We especially like bell beans because of their powerful, deep taproots that penetrate hardpans and bring up nutrients from the lower soil layers. They and their oat companion are both killed by winter temperatures and the beds where they were planted are ready for early spring crops.

Buckwheat

Buckwheat (*Fagopyrum esculentum*) is a popular green manure workhorse because it will grow in sandy, acid, and poor soils. It was the first green manure we sowed on our present farm. It is the most effective cover crop for suppressing competing weed growth because its 3-inch (7.5 cm) leaves and rapid growth shades out competing weed germination. Late-summer-sown buckwheat (four weeks before first expected frost) will provide a dense layer to protect the soil overwinter after it freezes, with no chance of it becoming a weed by setting seeds. It is the most difficult of our favorite green manures to save seeds from economically with our old-time techniques, so we continue to purchase seed.

Forage Radish (aka, Oilseed Radish)

Forage radish (*Raphanus sativus*) is another fall-planted, taprooted, soil-compaction-breaking, and winter-killed addition to the list of useful green manures. It establishes quickly in cool weather and grows a reasonably large, daikon-shaped radish that scavenges nitrate, which may have leached beyond the rooting zone of other green manures. It is almost completely decomposed by spring. We avoid planting it before other brassicas. The roots of this plant exude biochemicals called glucosinolates that help suppress pests such as nematodes. We intentionally sow them where a field tomato crop will be growing the following year.

Forage radish.

Mighty Mustard

Mighty Mustard is a trademarked type of mustard (*Brassica juncea*) produced by a farmers cooperative based in Idaho, and there are several varieties. This brassica-family green manure can serve as a soil biofumigant because it also contains glucosinolates. We use it when preparing a long-term grass pasture to become vegetable land again. We till the field first, then plant a buckwheat green manure for weed control. Then, after the buckwheat, we sow this mustard in late summer to knock back the wireworms (always a problem when coming out of a grass–legume pasture, especially before carrot or potato crops). We mow the mustard field with a flail mower at flowering stage just before seed set and incorporate the residues immediately to optimize the effect of the glucosinolates in discouraging the survival of the wireworms.

Red Clover

Red clover (*Trifolium pratense*) is a classic green manure and pasture plant. If I had no other legume for soil improvement, I would happily use it in all cases. We keep it on hand, mixed half and half with annual ryegrass (*Lolium multiflorum*) to use as a quick cover whenever we have nothing more specific in mind. Red clover is very deep rooting, and it complements the extensive fine root system of the annual ryegrass.

Forage Chicory

The untraditional inspiration to use forage chicory (*Cichorium intybus* L.) as one of our green manures came to me over 20 years ago, during a visit to organic vegetable farms in France. In addition to many other vegetables, all those farms were growing Belgian endive to store temporarily under cool conditions, and then to bring out in winter to sprout the white chicons called witloof chicory that are a popular winter salad in France.

When I noticed beautiful and pest-free winter crops on all these farms, I always asked about their rotation, specifically, what was the preceding crop. The answer was consistently Belgian endive. The popular pasture crop, forage chicory, is a close relative. As a further benefit, forage chicory is reported to accumulate quantities of sulfur, boron, molybdenum, and zinc, which are then made available to other crops when it decomposes. Back when we grew a lot of Belgian endive, these chicons were very popular at our winter farmers market, and we benefited from having an endive in the crop rotation. Radicchio, escarole, frisée, and curly endive are also members of the chicory family.

Some Odd Green Manures

If you happen to have a large patch of stinging nettle on your farm, I suggest mowing it with a scythe, and then rather than adding it to your compost, using it as a cut-and-carry green manure and spreading it directly in one of your fields. This was highly recommended to me by market gardening friends in Europe for nitrogen-loving crops.

Korean research found that turning under the remains of a spinach crop enhanced beneficial fungal microbes in the soil, which greatly increased the yield of a subsequent pepper crop. I suspect that in the future, growers will discover a number of other similar situations that may suggest modifying crop rotation sequences.

There are occasional opportunities to combine the benefits of green manures and crop rotation together. When researching crop rotations years ago, we learned that a preceding crop of soybeans before potatoes could help to control potato scab, and that a preceding crop of sweet corn gave the highest yields of potatoes. We already had the potatoes following sweet corn in our rotation, so, after two cultivations, we undersow our sweet corn to two rows of forage soybeans once the cornstalks are 10 to 12 inches (25 to 30 cm) tall (to give the corn a head start). The soybean plants grew tall enough to give very effective weed control in the corn field, and we gained the best of both effects.

In his delightful book *Weeds: Guardians of the Soil* (1950), Joseph Cocannouer philosophizes on how learning to use and manage the growth of weeds for many of their unrecognized benefits may be a wiser approach than simple eradication. The

Undersown forage soybeans can provide excellent weed control in sweet corn, as shown here.

one weed we have more or less succeeded in using that way is galinsoga (*Galinsoga parviflora*) because of its extensive, soil-filling root system. Whenever it appears in moderate amounts, we tolerate it. When we see the first small yellow blossoms we hoe or cultivate immediately before it sets seed. If I ever find another green manure crop with an equally indomitable root system, I will happily include it.

I am aware that several popular green manure varieties are not listed here. I only have experience with those I've mentioned. They were chosen for their dependability in this climate and for specific virtues, such as weed suppression or very deep roots, plus our ability to easily save seed from most of them. Although we could add more, the dozen or so species we presently use are sufficiently varied to satisfy my desire for the extensive biological synergy that comes from the presence of many and varied plant residues decomposing in the soil. Should I discover a need for more variability, it could be easily met by purchasing a few more bags of green manure seeds.

It is logical to ask what might be *missing* in a soil fed only with green manures and cover crops grown on site. Obviously, if some element needed for healthy plant growth is not present in the soil, it will not be available to the plants. That is why at Four Season Farm we have paid close attention over the years to supplementing our native soil with unpolluted, broad spectrum, long-lasting, finely ground, natural rock particles that contain everything the earth contains, as a low-solubility mineral reserve. If some mineral is not present in that source, I would not know how to find it.

I am always concerned about what pollutants might be brought in from outside the farm. The wise course of action, in my opinion, is to limit the materials you use to supplement your

Some growers use sunflowers as a green manure crop.

soil. You can't go wrong by using the remains of deep-rooted green manure plants and crop residues grown right on your soil. Ground particles of rocks laid down eons ago, and of known composition, are also a safe choice. Whereas in the past, we might have incorporated small amounts of seaweed from our local beaches, we now depend solely upon the homegrown soil fertility that we are able to create right here on the farm in order to faithfully provide our customers with clean produce. Sad to say, but organic wastes brought in from outside the farm, whether acquired free locally or purchased, always run the risk of inadvertently introducing into the soil undesirable industrial chemical residues from our increasingly polluted planet. A particularly tragic example is the recently publicized closing of several Maine farms, because their soils are permanently contaminated with per- and polyfluoroalkyl substances (PFAS)—synthetic chemicals—due to past state encouragement to dispose of sewage sludge by spreading it on farmland.

CHAPTER SEVEN

How All This Works

Two classic old-time practices are interwoven in our self-fed farm project: crop rotation and green manures. So let's start with the popular crops that we grow every year at Four Season Farm. Those 40 to 50 different crops belong to 10 botanical families which I categorize informally into families: grain, onion, beet, squash, cabbage, pea, carrot, tomato, and lettuce, along with herbs, which consist of basil for our field-grown crops. (We grow other popular herbs in a separate herb garden near the farm stand.) There are proper Latin names for each of these botanical families, but I prefer to keep it simple here.

The first general rule of crop rotation is to not grow the same crop or a closely related crop in the same spot in successive years. That should be easy enough to do if you consult the "Crops by Family" lists on page 124 to plan out a crop rotation sequence on the fields of your farm. Things get slightly more complicated when we add in the green manures: four of them are members of the pea family (clover, vetch, alfalfa, pea), two of them are grains like corn (rye, oats), and one of them, the forage radish, belongs to the cabbage family. Buckwheat is not a grain and is not related to any major vegetable.

Crops by Family

Grain Family	Onion Family	Beet Family	Squash Family	Cabbage Family
Sweet corn	Garlic	Beets	Cucumbers	Arugula
	Leeks	Chard	Melons	Asian greens
	Onions	Spinach	Summer squashes	Broccoli
	Scallions		Winter squashes	Brussels sprouts
	Shallots			Cabbage
				Cauliflower
				Kale
				Radishes
				Rutabaga
				Turnip

The usable farmland at Four Season Farm (approximately 3 acres) consists of 75 "fields," each 30 feet by 50 feet (9 by 15 m). One-third of the farm, or twenty-five of the fields, is used each year to grow late-harvest storage crops such as potatoes, squash, carrots, beets, cabbage, celeriac, and turnips, plus other fall crops such as broccoli, cauliflower, brussels sprouts, kale, spinach, and leeks. As was mentioned in chapter 5, by late October most all those fields have been harvested and resown to rye and vetch green manure.

Seven of our fields are each covered by one of our seven 30-foot-wide by 50-foot-long moveable greenhouses. Depending on the greenhouse model and where the greenhouse is located, sometimes we attach wheels and, with the help of a hand winch, can move a greenhouse from one field to another. Or using the

Pea Family	Carrot Family	Tomato Family	Lettuce Family	Herbs
Beans	Carrots	Eggplant	Artichokes	Basil
Peas	Celeriac	Peppers	Lettuce	
	Celery	Potatoes	Radicchio	
	Fennel	Tomatoes		
	Parsnip			

tractor, we tow the greenhouse like a sled. Before we move the greenhouse to a new field, we first grow a green manure crop on that field and incorporate it into the soil. After we move the greenhouse, we plant the recharged soil with new crops that will grow under cover of the greenhouse. Once a previously greenhouse-occupied field has been uncovered, it can be cleaned up with the tiller and then sown to a green manure appropriate for preceding whatever vegetable crop will follow next in the rotation. In this way, even our greenhouse soil fertility is continuously maintained with green manures without needing to grow those green manures under cover of the greenhouses themselves.

When in use for growing vegetable crops, all fields are subdivided into eight 30-inch-wide (75 cm) growing beds separated by 12-inch (30 cm) paths (all beds span the full 50-foot length of a

First, we move a greenhouse over beds of newly planted vegetable crops (*top*). We till the newly uncovered plot and broadcast a green manure crop (*bottom*).

field). After tilling under the preceding green manure crops, we mark out the beds by driving back and forth across a field with the wheels on my John Deere 870 tractor set to 42 inches (1 m) on center. Positioning the tractor one bed width over at each pass leaves a 30-inch-wide (90 cm) bed delineated between the 12-inch (30 cm) wheel tracks. We try to plant or transplant all eight beds on the same day, if possible, with either the same or similar crops to keep irrigation efficient both for the crops and for the subsequent green manure seedings.

A vigorous stand of peas and oats.

Once we harvest those crops, we till in the vegetable residues. Ideally, we then wait two weeks to allow for digestion of the plant material, and then sow a green manure crop. The goal is to minimize the amount of time without growing crops on the soil.

We sow the green manure varieties that have small seeds on a bed-by-bed basis, by drilling the seeds with either the Four-Row Pinpoint Seeder or the Six-Row Seeder (both from Johnny's Selected Seeds). For the larger seeded varieties, we use our 30-inch-wide, hand-pushed, homemade seed drill (described later in this chapter on page 142). A drill-type seeder that can place the seeds directly in the soil at the right depth is always the best sowing method to ensure that the green manure seeds will all germinate successfully. For seeding larger areas, we broadcast the seeds onto the surface of the soil with a chest seeder, then go over the area a second time to press the seeds into the soil with a cultipacker pulled by the tractor.

The decision about which crop follows another will be determined by the crop rotation plan. The decision about which green manure to plant will depend upon the time of year, the soil conditions, and which crop will grow next on that area. An expected future weed problem may suggest sowing buckwheat; or a hunch that soil compaction may have affected the previous crop might suggest one of the deep rooting green manures like forage chicory, bell beans, or forage radish. The beauty of using green manures to maintain soil fertility is that they can be adapted to the specific needs of the moment.

We are continually reevaluating the many details of our self-reliant system by asking questions, such as:

- Which deep-rooting green manures will be most effective at extracting specific major and minor

Broadcasting pea, oat, and vetch seed using a chest seeder.

nutrients from as far down in the soil profile as their roots can go? (From our experience they are sweet clover, forage chicory, bell beans, and forage radish.)

- Which are the best overwintering legume green manures to plant in fall to be tilled in before planting nitrogen-loving crops the following summer? (Red clover and hairy vetch are excellent choices.)
- Which crop rotation sequences may have biologically synergistic benefits on subsequent crops? (This question is inspired by a 1952 study in New Jersey, which found that "land resting" in an unharvested sod crop for one year in every three years increased the productivity of the following vegetable crops enough to justify the practice.)[1]

Soil-Feeding Crops versus Animal Manure or Compost

How do soil-feeding crops (perhaps a more accurately descriptive name than green manure crops) compare with spreading animal manure or compost? Let's start with manure. Many sources suggest that green manure crops should be used like forages and grazed by cattle or sheep to turn them into animal manure. Fair enough. However, incorporating cattle or sheep into a simple vegetable farm means adding a whole livestock component. You must provide housing, fencing, and year-round care for the critters. You must acquire feed during winter or other times when green manure grazing might not be available. And then there's marketing. Vegetables you have grown can be harvested and sold with few complications. Livestock, on the other hand, must be slaughtered and butchered in government-approved facilities before they can be legally sold for food. In many parts of the country, they must be transported a good distance to one of those facilities. All those extra steps are needed just to transform green vegetable matter into a more concentrated form (from which nutrients have already been removed to nourish the livestock), when it could have simply been tilled-in while in its green vegetable state with much less effort and expense. Unless you have access to a guaranteed clean manure supply from outside your farm (as we did in our first year here, almost 60 years ago) you are looking at greatly complicating your life by adding four-legged livestock.

Compost is another excellent soil-improving product. The story of compost is similar to that of livestock manure, in that compost doesn't make itself. The materials for compost must be collected, transported, piled up, and managed. That means a lot of material handling of crop residues that could instead have been directly incorporated into the soil with the tiller right after the crops were harvested. Compost merely decomposes organic matter the farm already has, whereas green manures add quantities of additional organic matter. We do make compost on Four Season Farm but only with residues that can't be tilled in directly. Those residues are limited because we avoid the work of initially taking crops to a packing shed to be processed.

On harvest days we use a 30-inch-wide wheeled cart with a wooden work surface upon which we can process crops in the field. Take leeks for example. We pull the leeks in the field, cut off the excess tops and peel off any discolored outer layers on the shank, trim the roots, and leave all those residues in the field as mulch for the remaining leeks. Only as a final step do we take the almost fully prepped leeks to the packing room to be washed before crating. Once a whole leek bed has been har-

Stages of the process: A crop of leeks almost ready to harvest, with a green manure crop in foreground and beds with chopped plant residues in background.

vested, we till all those residues in directly prior to sowing a green manure crop.

That same technique of prepping crops in the field while harvesting, then leaving the residues to be tilled in later, is used with almost all our 50-some crops. Exceptions are trellised greenhouse tomatoes and cucumbers, whose vines remain in production until the end of their season, when we pull those vines and take them to the compost area; root crops like beets, carrots, and radishes, which often keep their tops and ideally need to be washed right after harvesting them; and any other crops that require further curing and processing, such as onions, garlic, and shallots. (We spread our homemade compost, when mature, to supplement the soil in our one nonmobile greenhouse.)

Our system of prepping crops in the field as part of the harvest has resulted in the least extra amount of work for us; it combines efficient harvesting with perpetual soil improvement. However, that efficient soil improvement only became possible after we expanded our original 2½ acres (10,120 sq m) of de-rocked vegetable land to 3 acres (12,140 sq m) by de-rocking a half-acre (2,020 sq m) of one of our hayfields and adding it in. That expansion required the same enormous amount of effort we had expended on the first 2½ acres to remove many large rocks. We used a Graham-Hoeme chisel plow to bring numerous hidden rocks to the surface, then rolled them into the tractor bucket for removal on the next pass.

Before adding that extra half-acre, we had always needed to instantly reuse a just-harvested area by immediately replanting to a new vegetable crop so our limited soil could always be in use. We had been substituting our labor for

Prepping a crop of cured onions in the field as we gather them.

adequate land. By that I mean once a bed had been harvested, we would pull the remaining crop roots and residues by hand, take them to the compost area, spread a new layer of compost, and make the soil ready for seeding or transplanting another vegetable crop. Once we included the extra half-acre, we were able to substitute that land for our previous labor. Because we no longer needed to put a particular field back into vegetables right away, we could quickly till under the old crop residues (rather than clearing them away by hand), seed a desired green manure, and leave the land to look after itself until we needed it again. We avoided all the work involved with compost making—cleaning up the area, transporting the residues, turning the compost, and so forth—by leaving that identical amount of organic matter that we would have taken to the compost area to decompose in the soil where it had grown. We further avoided the labor of delivering a layer of finished compost. By the time we need that same patch of soil again, the tilled-in organic residues will have decomposed into reusable nutrients and a successful green manure crop will have added far more fertility than those residues would have by themselves, had they been composted.

Can Green Manures Outcompete Weeds?

Weed growth inhibits crop growth by competing for sunlight, moisture, and nutrients. Growing green manures and cover crops can, in turn, inhibit weeds by using those same competitive processes against them.

The classic green manure for quick growth and shading out weeds is buckwheat. It is easy to sow with a chest seeder. It is also easy to see the individual seeds lying on the soil after sowing, meaning you can quickly evaluate how densely you have sown it.

By early August, a stand of our rye–vetch mixture is flowering and has reached chest height.

This hand-pushed seed drill that Eliot made works well for sowing green manures that have larger seed.

Whenever shading and smothering weeds is your goal, it is wise to use a heavier seeding rate. For starters you might want to try three times the recommended 1 pound (.5 kg) per 500 square feet (46.5 sq m) and sow 3 pounds (1.4 kg). However, if you are drilling the seeds (better germination) instead of broadcasting them, then 2 pounds (0.9 kg) may be sufficient.

The winter rye and vetch we sow in the preceding fall attains a height of 6 feet (1.8 m) by early August, and it is obviously an effective shader and smotherer of weed crops. We mow that field then, as mentioned in chapter 5, with a flail mower, minus the back plate and depth roller. And we back into the crop so that the straw will get thoroughly chopped rather than pushed over. The 2 inches or so of chopped rye and vetch straw mulch covering the soil following that mowing acts as an initial weed deterrent. The large number of grains yielded by that mature rye and vetch, now lying on the soil under that mulch at about 20 times the standard sowing rate, certainly fulfils the heavy green manure sowing advice given earlier. Once all those rye and vetch seeds germinate and grow up through the mulch, there is little likelihood of any weed seeds germinating. If any do appear between the August mowing and the early April tilling the following spring, they can be easily discouraged by topping that area with the flail mower. When topping a green manure sowing to prevent the seeding of any weed varieties that have grown taller than the green manure, it is important not to cut so low that green manure regrowth is inhibited.

Consider also the stale seedbed weed control technique. This technique is highly recommended to vegetable growers in order to avoid weeds right from the start. It can be achieved by waiting for weeds to emerge and then shallowly cultivating, or flame weeding, the prepared area a few times so as to create a clean as

Mulching with straw in a home garden keeps down weeds and also helps to regenerate soil organic matter.

possible bed before new sowings or new transplants. The best green manures to use for getting the drop on new weed growth are those with large, quick-germinating seeds, such as peas and oats. If the season is too late for a buckwheat crop, we often use a pea and oat combination (due to its fast germination) in instances when we are aware that, for whatever reason, weeds might have gone to seed in a previous crop.

If you are presently achieving a less than ideal complete soil cover from a surface-sown green manure, despite using a cultipacker to establish the seeds, then you may be interested in a seed drill to be sure the seeds are sown at an ideal depth for better germination. When I was doing green manure trials (both standard and undersown) on a farm in Vermont in the 1980s, I built my own hand-pushed seed drill to use in the trial plots. I bolted together five EarthWay seeders rigged with a common push bar and a common front axle, so all five seeders would drop seeds simultaneously. Nowadays multiple-row Jang seeders are available, but they would cost a lot more and probably not work any better than my simple homemade solution. The only complication in creating that five-row seed drill was the arms that hold the soil covering chains on the EarthWay seeders, which stick out to the side. Those needed to be bent down to allow the five seeders to fit closely side by side. No problem with that. The original prototype was built 30 inches (75 cm) wide to fit between the rows of many crops. I suspect you could expand that to 48 inches (1.2 m) for larger areas and still be able to push it by hand.

All of these tools and techniques have a reason. Someone wishing to copy our green manure solution to bounteous soil fertility for vegetable crops needs to carefully consider all the various factors involved in the successful sowing, establishing, and incorporating of green manures to make sure their system is working as well as ours does.

Making This Work in the Home Garden

On December 15, 2024, my 86th birthday, Barbara and I retired from daily market garden work and leased Four Season Farm to an energetic young couple who are doing a great job maintaining our quality standards and thereby keeping our long-time customers happy. After many years of commercial market gardening, we were ready for a break. We were looking forward to resuscitating our home garden, which is about 50 feet by 50 feet (15 by 15 m) in size, but we wanted its care to be less intensive and less complicated than the farm had been. So we set up our home garden in the following way.

In our home garden, we plant succession crops after we harvest early crops. Here, we've planted young brassicas where early peas had grown.

First, we divided the garden space in two halves. We grow vegetables on one half and we sow the other half to a grass-legume pasture mix to care for the soil and store up fertility for bounteous vegetables the following year after that sod crop has been tilled in. The two halves of the garden change roles every year.

In the vegetable half, we plant in the spring as usual. As the growing season progresses, we try to be productive by replanting (for example, we follow early peas with brassica transplants for later harvest). The only green manure we use is buckwheat for temporary weed control following, say, an early crop like spinach and as a placeholder that helps prevent weed growth and keeps living roots in the soil to feed the soil microbes until we need that space again for a different vegetable crop. Gardeners worry about buckwheat going to seed and becoming a weed problem in its own right. But buckwheat won't go to seed if you mow it at just the right time. Set your rotary lawnmower on high and mow before the buckwheat is 20 percent in flower. The plants will regrow but stay vegetative. By using that technique, we can let numerous sowings of buckwheat continue to do their weed-smothering job until the end of the garden season. Then we leave the buckwheat to freeze in place and protect the soil over winter.

Early the next spring we rake off the buckwheat residue and add it to the compost heap. Then, we sow that half of the garden, used the previous year to grow vegetables, with the perfect "rest and rejuvenation crop"—a grass-and-legume pasture mix that is sown in the same way as sprinkling grass seed to plant a lawn. After broadcasting the seed, we mix it into the surface of the soil with a stiff-tined wire rake. We roll the area to press the seeds in, irrigate, and then leave the area alone all summer to grow guaranteed fertility for next year's garden. Some weed seeds from the soil seed bank may germinate along with the pasture mix. Any weeds that manage to thrive and stick up through the pasture mix can be easily controlled by topping them with a high-set lawn mower before they can make seed. The following spring, we till in that pasture sod as early as possible to make the area ready for planting vegetables. It may be tempting to till it the previous fall, but studies show that fall-tilled green manures can lose a large percentage of their stockpiled fertility over winter. However, if you pay close attention to weather reports in late fall, the green manure can be tilled in just before the soil becomes frozen for the winter, meaning the losses from fall rains should be very minimal.

We have chosen to sow the 8094 Pasture Mix from Fedco Seeds because it offers a good mix of grasses along with 10 percent pre-inoculated Rivendell white clover. I suspect you won't go wrong if you visit the seed-and-feed store in your hometown and buy what they are selling as a locally adapted grass–legume pasture mix. (Look back at the comments by Peter Henderson and John E. Weaver on page 52, about the soil-fertility-enhancing value of sod, and consult the "land resting" observations, page 132, from the New Jersey research from 1952.) All the soil fertility authorities mention the almost virgin-like soil that follows a year in sod. As for us, this work-saving solution is perfect for enhancing soil fertility in our divided-in-half home garden.

CHAPTER EIGHT

L'Envoi

It is comforting these days for me to look out the window any time of the year and see that every field on Four Season Farm is either growing a seasonal vegetable crop to be sold at one of our markets or is sown to a protective green manure crop under which it is gaining fertility to grow future crops. There is never an unproductive piece of ground on the self-fed farm or garden. Furthermore, there has been no need for me to worry about acquiring or affording purchased compost or manure supplies, because they are not components of this system. Relying on homegrown soil fertility keeps my farmer's mind steadily focused on optimizing the numerous natural soil processes involved in the creation of that fertility. And those processes guarantee the purity of my crops because no potentially contaminated inputs from off the farm are involved. In addition, and best of all, some part of the land, under our system of sequential green manures, is always ready for planting the next crop.

We have only 3 acres (12,140 sq m) to look out upon, but that same satisfaction could be experienced on almost any size of market garden. Green manures are a scale-neutral soil improver, meaning they work on any scale. In fact, a much larger farm than this might benefit by possibly having access to a small combine for harvesting seeds and in that way grow all their own seed for sowing green manures and make this dependable system even more self-reliant.

Our big biomass rye and vetch planting every three years is our answer to making sure we incorporate enough mature high lignin residues, such as tall rye and vetch straw, to continue increasing our total soil organic matter content. There are occasional suggestions in some of the studies cited in the book *Green Manuring* (1927) by Adrian Pieters that an annual green manure program may not be sufficient to fully maintain soil organic matter year after year. But by including our 19-month rye and vetch combination in our green manure sequence, we overcome that possible weakness in a way that fits in with our crop rotation and land use realities.

On the self-fed farm, you are in charge. You choose what green manures to grow and for how long. You grow them right where you want them and, if timing or season of the year are inappropriate for your first choice, you have a wide selection of other options. All of this is taking place right in your own increasingly fertile soil. You are the one determining your soil's future productivity as long as you are willing to pay attention to the important daily decisions about sowing and managing green manure crops with which you did not previously need to concern yourself. You have replaced all the trouble and expense of acquiring and spreading purchased manures or composts or organic plant stimulants with the daily attention to detail required to grow your own guaranteed-unpolluted soil fertility. I consider that a very logical step toward truly dependable organic vegetable production for the twenty-first century.

In recent years organic vegetable production has become overly focused on importing organic wastes to increase soil

organic matter, with little thought or understanding given to how the quality of that imported organic matter might affect the quality of the food produced. The information in this book is an attempt to return "organic," as practiced today, to its purer, less industrial roots. Organic food production is a living system in every way. Grown-on-site green manures are the logical answer to reestablish the creation and maintenance of soil fertility as part of that living system.

APPENDIX

Green Manure and Crop Sequences at a Glance

Here is a summary of possible sequences of green manure crops and vegetable crops, based on past experience at Four Season Farm and my calculated hunches.

Green Manure	When to Sow
Pea/oat	Late summer
Pea/oat	Early spring
Oats	Fall
Rye/vetch	August to no later than mid-October
Vetch	Fall
Sweet clover	Late winter snow seeding
Bell beans/oats	Late summer
Forage radish	Early fall
Buckwheat	Late summer
Red clover/ ryegrass	As desired
Mighty Mustard	Late summer (after a buckwheat crop)
Soybeans undersown in sweet corn	Corn in late spring; soybeans undersown in early summer
Forage chicory	Summer

When to incorporate	Crops to Follow
Early spring (till or rake aside)	Most early field crops
Midsummer	Winter carrots (sown August 15)
Winter-killed residue raked aside in early spring	Peas in spring
April (about 19 months after sowing)	Potatoes, field tomatoes
April	Sweet corn
Mid-July	Fall brassicas, late sweet corn
Early spring	Early spinach, lettuce, carrots, beets, etc.
Early spring	Field tomatoes
Early spring	Onion family crops
Before seeds set	Universal
Fall. Mow before seed set. Incorporate immediately.	Carrots and potatoes after pasture
Fall	Potatoes in spring
Spring or fall	Field tomatoes, fall brassicas

NOTES

Chapter 1. Celebrating Life

1. George Vivian Poore, *Essays on Rural Hygiene* (Longmans, Green and Co., 1893), 54, 85, 86, 89, 96.
2. Lord Northbourne, *Look to the Land* (J. M. Dent and Sons Ltd., 1940), 97, 99, 105.

Chapter 2. The Start of Purchased-Input Farming

1. Steven Stoll, *Larding the Lean Earth* (Hill and Wang, 2002), 194.
2. Jimmy M. Skaggs, *The Great Guano Rush* (St. Martin's Griffin, 1994). Under the authority of the Guano Islands Act of 1856, the United States eventually gained control of 94 islands in the Pacific and the Caribbean. As of 2025, 10 of those islands are still claimed under the act, although it does not appear that any guano is being mined on them presently.
3. Stoll, *Larding the Lean Earth*, 189–90.
4. Edward Faulkner, *A Second Look* (University of Oklahoma Press, 1947), 10, 52.
5. Leonard Wickenden, *Gardening with Nature* (Devin-Adair Company, 1958), 8, 251.
6. M. M. Morland et al., "Clay Amendments on Sand and Organic Soils," *Michigan State Quarterly Bulletin* 40, no. 1

(1957): 23–30; Gerhard Reuter, "Improvement of Sandy Soils by Clay-Substrate Application," *Applied Clay Science* 9 (1994): 107–20.

Chapter 3. The Quest for a Permanent Agriculture

1. Robert H. Elliot, *The Clifton Park System of Farming* (Faber and Faber, 1944), 61.
2. Peter Henderson, *Garden and Farm Topics* (Peter Henderson & Co., 1884), 193–94.
3. John E. Weaver, *Root Development of Field Crops* (McGraw-Hill Book Company, 1926), 204–5.
4. Robert H. Elliot, *The Agricultural Changes Required by the Times, and Laying Down Land to Grass* (J & J. H. Rutherfurd, Kelso, 1898), 7.
5. Elliot, *The Clifton Park System of Farming*, 44.
6. Sir R. George Stapledon, *Ley Farming* (Faber and Faber, 1948), 33, 35.
7. Cyril G. Hopkins, *Soil Fertility and Permanent Agriculture* (Ginn and Company, 1910).
8. Cyril G. Hopkins, "Shall We Use 'Complete' Commercial Fertilizers in the Corn Belt?" *University of Illinois AES Circular* 165 (1912): 1–20.
9. USDA, *Soils and Men* (US Government Printing Office, 1938), 348–49.
10. USDA, *Soils and Men*, 348–49.
11. Louis Bromfield, *From My Experience* (Harper and Brothers, 1955), 309.

Chapter 4. In Defense of Intelligent Tillage

1. CSANR, "Without Big Biomass, Green Manures Are a Step Backwards," CSANR, Washington State University (2018).

2. J. N. Sorensen and K. Grevsen, "Strategies for Cut-and-Carry Green Manure Production," *Acta Horticulturae* 1137 (2016).
3. Section V, Soil Health & Land Management, of *The Framework for Regenerative Organic Certified*, Version 4.1, June 27, 2023.
4. Selman Waksman and Robert Starkey, *The Soil and the Microbe* (John Wiley & Sons, Inc., 1931), 216.
5. Edward Faulkner, *Soil Development* (University of Oklahoma Press, 1952), 231.
6. Harry Snyder, *Humus in Its Relation to Soil Fertility* (USDA Yearbook of Agriculture, 1896).
7. J. L. Hills, C. H. Jones, and C. Cutler, "Soil Deterioration and Soil Humus," Vermont Agricultural Experiment Station, Bulletin 135 (1908): 142–77, College of Agriculture, Burlington, Vermont.
8. W. D. Keller, "Native Rocks and Minerals as Fertilizers," *The Scientific Monthly*, 66 (February 1948): 122–30.
9. P. Swoboda et al., "Remineralizing Soils? The Agricultural Usage of Silicate Rock Powders. A Review," *Science of the Total Environment* (2021).

Chapter 5. Putting Green Manures at the Heart of the System

1. A CoolBot unit is an electronic controller that resets the thermostat of a conventional air conditioner to make it run much colder so it can be used to turn any insulated, confined space into a low-cost walk-in cooler.
2. Cyril G. Hopkins, "Shall We Use 'Complete' Commercial Fertilizers in the Corn Belt?" *University of Illinois AES Circular* 165 (1912): 1–20.

3. Peter Whoriskey, "The Labels Said 'Organic.' But These Massive Imports of Corn and Soybeans Weren't," *Washington Post* (May 12, 2017).

Chapter 7. How All This Works

1. O. R. Neal, "Effects of Land Resting on Conservation and Productivity of Vegetable-Growing Soils," *Agronomy Journal* 44 (1952): 362–64.

BIBLIOGRAPHY

Bailey, Liberty Hyde. *The Holy Earth*. Ithaca, NY: Comstock Publishing Co., 1919.

Bailey, Liberty Hyde. *The State and the Farmer.* Saint Paul: Minnesota Extension Service, 1996.

Beeman, Randal S., and Pritchard, James A. *A Green and Permanent Land*. Lawrence: University Press of Kansas, 2001.

Bromfield, Louis. *Pleasant Valley*. New York: Harper and Brothers, 1943.

Bromfield, Louis. *From My Experience*. New York: Harper, and Brothers, 1955.

Bunch, Roland. *Restoring the Soil.* Winnipeg: Canadian Foodgrains Bank, 2012.

Cocannouer, Joseph A. *Weeds: Guardians of the Soil.* New York: Devin Adair Company, 1950.

Coleman, Eliot. *The New Organic Grower.* White River Junction, VT: Chelsea Green Publishing, 2018.

Darwin, Charles. *Darwin on Humus and the Earthworms.* London: Faber and Faber, 1945.

Farb, Peter. *Living Earth*. New York: Harper & Brothers, 1959.

Elliot, Robert H. *Agricultural Changes Required by the Times.* Kelso, Scotland: J. and J. H. Rutherfurd, 1898.

Elliot, Robert H. *The Clifton Park System of Farming and Laying Down Land to Grass.* London: Faber and Faber, 1944.

Faulkner, Edward. *Plowman's Folly.* Norman: University of Oklahoma Press, 1943.

Faulkner, Edward. *A Second Look.* Norman: University of Oklahoma Press, 1947.

Faulkner, Edward. *Soil Development.* Norman: University of Oklahoma Press, 1952.

Harlan, Caleb. *Farming with Green Manures, on Plumgrove Farm.* Philadelphia: Alfred J. Ferris, 1899.

Henderson, Peter. *Garden and Farm Topics.* New York: Peter Henderson & Co., 1884.

Hopkins, Cyril. *Soil Fertility and Permanent Agriculture.* New York: Ginn & Company, 1910.

Klay, Ruedi, and Schmid, Otto. *Green Manuring.* Temple, ME: Woods End Agricultural Institute, 1984.

Krasil'nikov, N. A. *Soil Microorganisms and Higher Plants.* Washington, DC: National Science Foundation, 1961.

Little, Charles E. *Green Fields Forever.* Washington, DC: Island Press, 1987.

Magdoff, Fred, and van Es, Harold. *Building Soils for Better Crops,* 2nd edition. Beltsville, MD: Sustainable Agriculture Network, National Agriculture Library, 2000.

Maude, Hugh. *The Farm, a Living Organization.* Dundalk, Ireland: Dundalagan Press, 1943.

Nearing, Helen and Scott. *Living the Good Life.* New York: Schocken, 1970.

Northbourne, Lord. *Look to the Land.* London: Dent, 1940.

Picton, Lionel James. *Nutrition and the Soil.* New York: Devin-Adair, 1949.

Pieters, A. J. *Green Manuring.* New York: John Wiley & Sons, 1927.

Piper, C. V., and Pieters, A. J. *Green Manuring*, USDA Farmers' Bulletin No, 1250, 1925.

Poore, George Vivian. *Essays on Rural Hygiene.* London: Longmans, Green and Co., 1903.

SARE. *Managing Cover Crops Profitably.* Sustainable Agriculture Research and Education Program, 2007.

Sarrantonio, Marianne. *Northeast Cover Crop Handbook.* Emmaus, PA: Rodale Institute, 1994.

Skaggs, Jimmy. *The Great Guano Rush.* New York: St. Martin's Griffin, 1994.

Stapledon, Sir R. George. *Ley Farming.* London: Faber and Faber, 1948.

Stoll, Stephen. *Larding the Lean Earth.* New York: Hill and Wang, 2002.

USDA. *Soils and Men.* Washington, DC: US Government Printing Office, 1938.

Waksman, Selman. *Humus.* Baltimore: Williams and Wilkins Company, 1938.

Waksman, Selman. *Soil Microbiology.* New York: John Wiley & Sons, 1952

Waksman, Selman, and Starkey, Robert. *The Soil and the Microbe.* New York: John Wiley and Sons, Inc., 1931.

Weaver, John E. *Root Development of Field Crops.* New York: McGraw-Hill Book Company, 1926.

Weaver, John E., and Bruner, William E. *Root Development of Vegetable Crops.* New York: McGraw-Hill Book Company, 1927.

Wickenden, Leonard. *Make Friends with Your Land.* New York: Devin Adair Company, 1949.

Wickenden, Leonard. *Gardening with Nature.* New York: Devin Adair Company, 1958.

IMAGE CREDITS

The images on pages vi, viii, x, xiii, xvi, xxii, 3, 6, 8, 13, 14, 18, 21, 23, 25, 29, 30, 34–35, 39, 40, 42, 44, 49, 50, 53, 57, 58, 61, 66, 69, 72, 73, 80, 82, 83, 84, 87, 90–91, 92, 98, 100, 104, 108, 120, 122, 126–127, 128, 131, 134–135, 137, 139, 142, 144–145, 148, 151, and 156 are by Barbara Damrosch.

The images on pages 89, 94, 95, 118, and 140 are by Eliot Coleman.

The image on page 111 is from Sheryl / Adobe Stock Images.

The image on pages 76–77 is courtesy of Robbie George.

The images on pages 46–47, 112, 164, and 166 are courtesy of Clara Coleman.

The images on pages xix, 54, 115, and 129 are used courtesy of Johnny's Selected Seeds.

The image on page 182 is by Lynn Karlin.

INDEX

Index

D

Index

H

Index

R

Index

S

ABOUT THE AUTHOR

Eliot Coleman is one of the most influential thought leaders in the organic farming movement. He has over fifty years' experience in all aspects of organic farming, with special expertise in growing field vegetables and greenhouse vegetables. He is the author of *The New Organic Grower*, *Four-Season Harvest*, and *The Winter Harvest Handbook*. Coleman and his wife, Barbara Damrosch, have recently retired from active management of the commercial year-round market garden at Four Season Farm in Harborside, Maine, to focus on horticultural research projects, including the optimal use of green manures to replace outside inputs in organic growing systems. To learn more about Eliot and Barbara, visit their website, eliotbarbara.com.